THE PENGUIN

AND THE LEVIATHAN

Also by Yochai Benkler

The Wealth of Networks

THE PENGUIN AND THE LEVIATHAN

THE TRIUMPH OF COOPERATION OVER SELF-INTEREST

YOCHAI BENKLER

Published in the United States by Crown Business, an imprint of the Crown Publishing Group, a division of Random House, Inc., New York.

www.crownpublishing.com

Crown Business is a trademark and Crown and the Rising Sun colophon are registered trademarks of Random House, Inc.

Crown Business books are available at special discounts for bulk purchases for sales promotions or corporate use. Special editions, including personalized covers, excerpts of existing books, or books with corporate logos, can be created in large quantities for special needs. For more information, contact Premium Sales at (212)572-2232 or e-mail specialmarkets@randomhouse.com.

Library of Congress Cataloging-in-Publication Data
Benkler, Yochai.
The penguin and the Leviathan: the triumph of cooperation over self-interest/Yochai Benkler.
p. cm.
1. Cooperation. 2. Self-interest. 3. Altruism. 4. Identity (Philosophical concept) I. Title.
HD2963.B46 2011
306.3'4—dc22 2010038576

ISBN 978-0-385-52576-3
eISBN 978-0-307-59019-0

Printed in the United States of America

Book design by Leonard W. Henderson
Jacket design by Daniel Rembert

10 9 8 7 6 5 4 3 2 1

First Edition

For the millions who, by their acts every day,

small and large, give humanity its name.

CONTENTS

CHAPTER 1

THE PENGUIN VS. THE LEVIATHAN

What do Southwest Airlines or Toyota's shop-floor processes, Chicago's community policing program, and Wikipedia or Linux have in common? The answer is that they are all systems that have relied on human cooperation, rather than purely on incentive compensation, punishment, or hierarchical control. Toyota structured its shop-floor relations around teamwork, and engineered its supplier relations into a collaborative network built on trust and long-term cooperation, instead of a top-down system governed by process engineers and competitive bidding. Until its recent troubles, for more than two decades its model was seen as revolutionary and the primary reason for its becoming the world's largest automobile manufacturer. Southwest Airlines similarly has outperformed its competitors by a wide margin, fostering a collaborative team spirit based on relative autonomy, high trust, and a pervasive sense of fair treatment among employees. Rather than cracking down harder on offenses, the Chicago police force has, for two decades, led the drive to a community-based model of policing, in which neighbors and police officers work together to prevent crime more effectively. Wikipedia relies on content created entirely by volunteers, allowing anyone and everyone to contribute their time and knowledge, rather than paychecks or editorial control, and Linux, the open-source software, depends on a massive collaboration between volunteers and paid contributors, whose outputs they all share and no one owns exclusively.

The way these organizations work flies in the face of what has long been the dominant assumption in Western society about human motivation: that human beings are basically selfish creatures, driven by their own interests. For decades economists, politicians and legislators, business executives and engineers have acted as though all systems and organizations had to be built around incentives, rewards, and punishments in order to get people to achieve public, corporate, or community goals: If you wanted to reduce crime, enact stronger penalties, like the three-strikes laws in California that sentence people to prison for life if they are convicted of a third felony. If you want employees to work harder, incorporate pay for performance and monitor their results more closely. If you want executives to do what's right for shareholders, pay them in stock. If you want doctors to take better care of patients, threaten them with malpractice suits so that fear of litigation makes them take better care. The model in all these is the same—we are driven by self-interest; to get others to act well, you have to monitor, reward, and punish them. And yet all around us we see people cooperating and working in collaboration, doing the right thing, behaving fairly, acting generously, caring about their group or team, and trying to behave like decent people who reciprocate kindness with kindness. Nowhere has this fact been more obvious than online, where Wikipedia and open-source software have been so successful. Tux, the Linux Penguin, is beginning to nibble away at the grim view of humanity that breathed life into Thomas Hobbes's Leviathan.

• • •

Testimony of Dr. Alan Greenspan to the Senate Committee of Government Oversight and Reform,
OCTOBER 23, 2008

Alan Greenspan: Those of us who have looked to the self-interest of lending institutions to protect shareholder's equity (myself especially) are in a state of shocked disbelief.

Senator Henry Waxman: In other words you found that your view of the world, your ideology was not right. It was not working.

> **Greenspan:** Precisely . . . that's precisely the reason I was shocked because I've been going for forty years or more with very considerable evidence that it was working exceptionally well.

Former Federal Reserve chairman Alan Greenspan's unwavering belief in the power of self-interest is based on two of our society's most widely held, long-standing, and erroneous assumptions. The first is the assumption that inspired philosopher Thomas Hobbes's *Leviathan:* that humans are fundamentally and universally selfish, and the only way to deal with people is for governments to step in and control us so that we do not, in our shortsighted pursuit of self-interest, destroy one another (or make one another's lives too miserable to bear). The second assumption was Adam Smith's alternative solution to our assumed selfishness—the Invisible Hand. Smith's *Wealth of Nations* argued that because humans are inherently self-interested and human decision making is driven by the rational weighing of costs and benefits, our action in a free market would tend to serve the common good. In other words, in our pursuit of self-interest we would work to fulfill one another's needs, not because we care about one another's well-being, but because it is mutually advantageous to do so. Though their prescriptions are quite different, both Leviathan and the Invisible Hand have the same fundamental starting point: a belief in the selfishness of mankind. The former tries to curb and control selfish human behavior through monitoring and punishment; the latter imagines markets as places where self-interest will lead people to act in ways that serve the common good.

The major alternative in Western political thought coalesced from the work of a wide array of thinkers: from French philosopher Jean-Jacques Rousseau, through Scottish Enlightenment philosopher David Hume and Adam Smith's other major work, *The Theory of Moral Sentiments,* to the work of major anarchist philosophers Pierre-Joseph Proudhon and Peter Kropotkin. Broadly speaking, this far more flattering view sees human beings as fundamentally capable of empathy, of possessing sentiments that compel us to act morally, cooperatively, or generously, not only in our own self-interest. This does not claim us to be saints; it merely says that we are capable of virtue, and that we need not be robotic slaves to the government's Leviathan, automatons guided by the Invisible Hand of the market, or parts of the collectivist Hive of fascism to serve the common weal. In honor of Tux, the symbol of Linux, I'll call this alternative the Penguin.

Cycles of Leviathan and the Invisible Hand

Modern European and North American history has cycled between social, political, and economic systems that tended toward the Leviathan, and those that were based on the Invisible Hand. Throughout the seventeenth and eighteenth centuries, Europe's absolute monarchies were more or less inefficient versions of Leviathan (just substitute "monarchy" for "government"). The inefficiency in exercising control provided substantial breathing room for Invisible Hand and the social action, the Penguin, to flourish underneath the Leviathan, more or less informally. By the nineteenth century the decline of the monarchy, the Industrial Revolution, and the subsequent rise of commerce swept the Invisible Hand into power (nowhere with more painful effects than in Britain, as Friedrich Engels and Charles Dickens portrayed with such excruciating detail). This long reign of the Invisible Hand in both Europe and America was punctuated by panics and crashes throughout the nineteenth century, then, in 1929, came to a swift and abrupt end as the markets crashed, ushering in the Great Depression.

Now the pendulum swung violently in the other direction. In Germany, where industrialization had already suffered a major blow from World War I, and in Russia, where it had been passed over altogether, moving straight from the Czar's lethargic rule to Stalin's cruelly efficient model, Leviathan reared its ugly head with a viciousness unmatched before or since, in the form of fascism and Soviet communism. In the United States, Britain, and other liberal democracies, Leviathan took more benevolent forms: the burgeoning welfare state and the rise of government bureaucracies (ushered in by the New Deal in the United States, and by similar movements in Western Europe). By the late 1950s and early 1960s the pendulum began to swing back as concerns mounted over petty bureaucrats, unchecked discretion, and inefficiency. By the 1980s we were back in full swing toward laissez-faire capitalism; the Reagan and Thatcher governments in the United States and Britain, the rise of the efficiency- and free-trade-focused European Commission in Europe, and the emergence of the World Bank and the International Monetary Fund as bearers of what came to be known as "the Washington Consensus." The Invisible Hand seemed to have completely won when even the center-left governments of the United States and the United Kingdom, under Bill Clinton and Tony Blair, busied themselves with dismantling welfare as we know it—replacing government bureaucracies with privatized, market-based alternatives—and deregulating the financial markets that flourished in New York and London. The drive to weaken the state and make way for self-interest in the market reached new peaks under George W. Bush. Predictably, today we find ourselves facing a new crisis, our economic systems toppled by our blind faith in the power of self-interest and in our ability to harness it effectively through incentives and payoffs.

Where does this leave us? If neither the command-control systems dictated by the Leviathan nor the Invisible Hand of the free market can effectively govern society, where shall we turn? What, if anything, do systems based on cooperation have to offer besides the pleasant diversion or utopic ideal of a free operating system or a global online encyclopedia? Can the Penguin deliver us more robust, working social and economic systems that break us out of this vicious cycle?

I believe that he can.

Over the course of the twentieth century, intellectual trends in such diverse fields as business, anthropology, psychology, human evolution, economics, political science, and law have pondered the question, How shall we construct the systems we inhabit? We live our lives, after all, within the confines of systems: business systems, like workplaces and shopping malls; legal systems, like intellectual property laws or environmental regulations; technical systems, like the Internet, or the highways and bridges; administrative systems (some, like Medicare, run by the state; others, like arts and cultural foundations, by nonstate bodies); educational systems, like preschools and university research labs; and social systems, like our networks of friends.

Whether their goals were to increase profits, improve law and governance, advance the sciences, or simply help us to lead better, happier lives, leading researchers and thinkers have long sought to improve the way these systems are designed. In the twentieth century's first six decades, the favored approaches reflected the Leviathan; most systems were large, hierarchical, and controlled. Within the United States, this trend began in companies, when, early in the century, Frederick Taylor published his *Principles of Scientific Management,* which concerns a management process by which every action, by every employee, was described, timed, measured, and monitored to assure the most efficient operation, reducing the employee to a very well-regulated component in a perfectly designed system—one that was controlled by the powers above.

Henry Ford soon took this basic concept and embedded it in an assembly line. Eventually, this mentality—that workers were essentially robotic, mechanized creatures who could perform adequately at a given task only if properly monitored, managed, and supervised—permeated a broad swath of sectors and industries, from factory floors to boardrooms. This top-down hierarchy later expanded to the public realm, as New Deal administrative agencies were built on the Progressive Era assumption that expert agencies could plan more effectively than the irresponsible market that had led to the Crash of 1929. In Europe, the progression went in the other direction; the rise of command-control systems began with state bureaucracy (pioneered by Bismarck in Prussia) and later diffused to businesses. But whatever the order of events, by the mid-1960s one thing was clear: In the United States and elsewhere, organization of hierarchies had come to dominate modern economic and social life. The father of sociology, Max Weber, saw this earlier in the century; economists like Joseph Schumpeter saw this in the mid-century: The future was to be inherited by ever larger, controlled bureaucracies; by various versions of Leviathan.

Paralleling the arc taking place in politics, the intellectual debate (and to some extent the practice) of the next forty years saw a pronounced shift away from centralized systems and toward markets and market-mimicking approaches. In short, the Invisible Hand reemerged onto the landscape not only in the ivory tower and in the halls of Washington, but also in business and social life. Partly the shift was underwritten by the Cold War, the ideological battle between market-based and socialist economies. To a great extent, however, it was instigated by the inability of control-based systems to manage an increasingly complex and interconnected economy and society. As global trade expanded, and technological growth ushered countless new industries, companies, and products into the marketplace, hierarchical systems simply were inadequate to the task. Market-based systems, on the other hand, seemed not to require such close monitoring and control. By setting up systems of incentives, those who trusted in the efficacy of the market could let us run things more or less as we saw fit. This, it seemed at the time, appeared to be a far cheaper and more effective way of doing business, and it quickly caught on as a way of looking at the world. At the end of the twentieth century, the technology sector exploded, multiplying, several times over, the challenges of managing what was already a complex, far-flung and fast-moving world. Economists and businesses began to embrace a more starkly selfish model of humanity than Adam Smith had ever proposed, seeking more deregulation and relying, to an ever greater extent, on the perfectly aligned markets to harness our self-interest to the commonweal. We even began to accept the notion that our inherent selfishness applied not only to the business world or to the markets, but also to social life, to love and family. Economics was by now the unquestioned queen of the social sciences, and we began to think of human behavior in terms of almost mechanical, predictable, responses to punishments and incentives.

• • •

But in the last decade (though the roots of the current trend run back at least to the early 1980s), a series of changes has triggered a fundamental shift away from this theory of universal selfishness. First of all, businesses slowly began to learn the lessons that began with Toyota's productivity and quality improvements in its U.S. plants since the 1980s, relative to its American counterparts; the high-tech industry, personified in some sense in Google's playground image, suggested that relative informality and an emphasis on autonomy and creativity, as well as social engagement, were critical. More business school courses and businesses themselves began to emphasize and experiment with organizational models that were neither strictly market-based nor as hierarchical as had been thought necessary in the past. Instead they were built around the assumption that, given the right conditions, people would opt to cooperate and collaborate to serve the collective good of the organization—of their own free will. More radical still, the rise of peer production on the Net—from free and open-source software, to Wikipedia, to collaborative citizen journalism on sites like Daily Kos or Newsvine, to social networks like Facebook and Twitter—produced a culture of cooperation that was widely thought impossible a mere five or ten years ago. These changes did not happen at the fringes of society; they arose precisely in those places, like Silicon Valley, that represented the cutting edge of our social and economic trends. The business world finally began to sit up and take notice. Sites that depend on volunteer contributions became enormously popular (Wikipedia is the seventh or eighth most trafficked site on the web, with more than 300 million unique views worldwide per month). Businesses, some traditional, like IBM, others new, like Google or Facebook, Red Hat or Craigslist, now had the technology to experiment with these new models, and in turn found new ways of being profitable by *engaging* people, rather than controlling them. And, in the wake of the 2008 economic collapse, even some of the staunchest proponents of pure, free-market libertarianism had to acknowledge the limitations of a market-incentives-above-all model.

This shift toward a more optimistic, human, and humane view—that we as individuals can be motivated to productive ends by engaging one another socially and creating collaborative relationships—extends far beyond the business world or the networked environment. The success of collaborative efforts like Wikipedia and Linux has been paralleled in social systems as wide-ranging as the community policing modeled in Chicago (which has since been so widely and enthusiastically adopted elsewhere that by 1999, community policing efforts were under way in more than three-quarters of communities with over fifty thousand residents throughout the United States) to the CIA's interagency online collaboration platform, Intellipedia.

This adoption of cooperative systems in so many fields has been paralleled by a renewed interest among researchers in the social and behavioral sciences in the mechanics of cooperation. Perhaps humankind might not be so inherently selfish after all. Through the work of hundreds of scientists, we have begun to see mounting evidence in psychology, organizational sociology, political science, experimental economics, and elsewhere that people are in fact more cooperative and selfless, or at least behave far less selfishly, than most economists and others previously assumed. This isn't just theory; dozens of field studies have identified cooperative systems, often more stable and effective than equivalent incentive-based ones. Even in the study of human biology, evolutionary biologists and psychologists are now finding neural and possibly genetic evidence of a human predisposition to cooperate. Though it may sound counterintuitive, there is much evidence that evolution may actually favor individuals (and societies that include these individuals) who are driven to cooperate with or help others, even at cost to themselves (I'll talk about this in greater detail in the next chapter).

In fact, in hundreds of studies, conducted in numerous disciplines across dozens of societies, a basic pattern emerges. In any given experiment, a large minority of people (about 30 percent) behave as though they really are selfish, as the mainstream commonly assumes. But here is the rub:

Fully half of all people systematically, significantly, and predictably behave cooperatively. Some of them cooperate conditionally—they treat kindness with kindness, and meanness with meanness. Others are unconditional cooperators, or altruists, who cooperate even when it comes at a personal cost. The point is, across a wide range of experiments, in widely diverse populations, one finding stands out:

> In practically no human society examined under controlled conditions have the majority of people consistently behaved selfishly.

That's all fine and good in the lab, you might be thinking. But what does it mean for us in day-to-day life? Quite a bit, actually. It means that our existing social and economic systems—from our hierarchical business models, to our punitive legal system, to our market-based approaches to education—are often designed with the wrong model of who we are, and why we do what we do. That we don't need systems that see individuals solely through the lens of self-interest, possessing only desires and preferences. Using control or carrots and sticks to motivate us isn't effective. To motivate people, we need systems that rely on engagement, communication, and a sense of common purpose and identity. In other words, organizations would be better off helping us to engage and embrace our collaborative, generous sentiments, rather than assuming we are driven by self-interest. As we will see, there are settings where trying to combine systems based on self-interest—such as material rewards or punishment—will backfire and lead to less productivity than an approach oriented solely toward social motivations.

Why Has the Myth of Self-Interest Persisted?

If we have all this evidence pointing to the power and promise of collaboration, why do I still find myself sitting in conference rooms, eyes glazing over, as I listen to speakers explain with charts and graphs why the success of the free Linux operating system is just a temporary market imperfection, a blip, a flash in the pan that will disappear as soon as a pricing system can be properly put in place? Or how, despite the fact that its proprietary competitor, Microsoft *Encarta,* recently went out of business, Wikipedia will never be as good as this or that new for-profit alternative? Or how, even in the wake of the global economic meltdown, the Invisible Hand of Wall Street is still superior to other alternatives?

Why *do* so many of us still cling to this grossly unflattering view of the human species as a selfish animal? Why do we persist even in the face of so much evidence to the contrary? Why do we assume the worst about humankind? I think there are four reasons: one, the fact that the assumption of human self-interest is partially correct; two, the historical moment at which the concept of selfishness and self-interest rose to prominence in our culture; three, our desire for simple, clear, elegant explanations about ourselves and the world in which we live (even if those simple explanations are wrong); and four, the mere force of habit and its ability to distort our perception and thinking. Let's look at each of these more closely.

Partial truth. One reason that the myth of universal selfishness persists is that it is not *wholly* wrong—only mostly so. We all have experienced moments of conflict, when we are torn between what is good for us and what is good for others. Many among us have on occasion caved to our own self-interest. We can recognize ourselves in the story of selfishness. Moreover, it can't be denied that a fairly significant percentage of people do at times behave selfishly, and so statistically, each of us interacts with selfish people about one-third of the time (usually with unpleasant results if we don't find ways to control or channel such self-interested behavior). So if we ourselves sometime behave selfishly, and the people we interact with act selfishly a good deal of the time, it's easy to make generalizations about human nature, and assume the experts are right about the notion of universal self-interest. But human self-interest is only partially true. Yes, some people are selfish, but few are selfish all of the time—and some of us do not act out of self-interest. These generalizations often fail to take into account all the times we have acted selflessly, and all the times we have been the beneficiaries of generosity from others. This makes for a more complex, nuanced picture of the world. We now have to contend with how we can take full advantage of our sociability without exposing ourselves to too great a risk of abuse.

History. The roots of our assumption about universal selfishness are old. Nonetheless, the assumption has only recently come to dominate our scientific theory of human behavior. The trend rose to prominence in the United States in the 1950s and gained traction over the course of the following three decades. In other words, the belief that human beings will respond only to the punishments and incentives of a free-market system grew popular during the Cold War—a time when the global power struggle between the former Soviet Union and the United States was couched as an ideological battle between capitalism and free enterprise, on the one hand, and socialism and collectivism on the other. It was a time when even the slightest murmur of dissent against the free-market ideology suggested that one was a traitor: a card-carrying communist siding with an insidious enemy that threatened to destroy the West. In the era of red scares, blacklists, McCarthyism, and the Rosenbergs, one needed very good reasons, not to mention a good deal of courage, to question the scientific merit of our assumptions about human motivation and action that were offered as the very foundations of capitalism. Through the time of the Evil Empire, critiques of this position persisted, but it was largely denied a place in the mainstream.

Simplicity. Human beings tend to seek out neat, simple explanations to help us make sense of our confusing and complex world—coherent stories that help us organize many different facts, ideas, and insights and help us to predict what will happen if we do X, or what we will find if we look under Y. Some work on cognitive psychology anchors this in "cognitive fluency"—our tendency to remember and hold on to things that are simple to understand and remember. We seem to have a strong preference for, and tend to easily accept, simple explanations that allow for simple solutions (for example, if the crops failed, God must be angry). Even in scientific theory, Einstein famously said: "Everything should be made as simple as possible, but not one bit simpler." A straightforward, uncomplicated theory of human nature that reduces our actions as simple, predictable responses to punishments and incentives and helps us explain away confusing and even disturbing behaviors is incredibly appealing and attractive to the human mind. Our actual lived experience is much more complex. As the influential social scientist Michael Polanyi put it, "we know more than we can tell." This tendency to simplify what we "can tell" relative to what we know leads us to build organizational strategies, laws, and technical systems based on models that are always a touch too simple, always a bit out of touch with what we intuitively know.

Habit. Almost two generations have now been educated and socialized to think in terms of universal selfishness. Sure, if you asked a large group of people whether they believed human beings to be fundamentally selfish creatures, they might hopefully cite Peace Corps volunteers in Africa, or the United Way, or Mother Teresa's work with the poor

in Calcutta. But when pressed, each
probably recognize that these except
the overall way of the world. One can't turn
open a newspaper without hearing or reading about
tives," "bonuses," "scores," and "benchmarks." In any given
situation, "what's in it for him/her/us?" becomes the first question on everyone's minds. And once we get in the habit of thinking of ourselves and our behavior in a given way, we tend to interpret all the evidence we encounter and collect so it fits with our preconceptions and assumptions, by discarding some bits and emphasizing others. Through sheer force of habit, our erroneous beliefs and ways of thinking about human nature become reinforced, and get more entrenched over time.

But let me tell you a secret. We've all known, intuitively, that we aren't really selfish and rational all the time. We teach our children to tell the difference between right and wrong, to be nice to others, to follow the Golden Rule. We have all loved another person, and made crazy, irrational sacrifices in the name of love. (Even economists. Really.) We've all given up our seat to an old lady or handicapped stranger on the train, held the elevator or door for the person behind us, put a dollar in the UNICEF jar. We've all done things because we knew intuitively that they were simply the right thing to do, not because they would bring us personal gain or profit. In other words, we've all acted in ways that prove we know, on some level, that we are more selfless than the reigning view of humanity would suggest.

So how do we account for these selfless actions that we perform every day? How do we explain our innate sense of empathy, fairness, or doing the right thing? One answer is that, far from being robots or selfish brutes, we are moral creatures, with moral codes that surpass and triumph over rational calculation and self-interest. After all, the vast majority of us would not think twice about ruining a good suit to save a baby drowning in a swimming pool; our conscience demands it. But not only are we driven to follow our own internal moral code, we are driven to follow our society's moral code as well. And in most cultures, both complex and simple, one of those values is to help and collaborate with others. As Nobel laureate Amartya Sen wrote in his 1977 challenge to economics, *Rational Fools,* we do in fact undertake commitments—to ourselves, to others, or to a set of beliefs or values. When we do so, these commitments exert a real force on our decisions, even if they run against our material interests.

The second reason we care so much about this code of behavior is that, quite simply, we are social beings. We often want to get along with others in our social environment, by conforming to the values of our particular culture. Sometimes we do so consciously (one could say selfishly) because it is expedient—we're trying to ingratiate ourselves, or we're trying to be seen as kind or thoughtful. Other times we are not sure what is socially appropriate, so we instinctively take cues from people we see around us. But often we follow our cultural mores and norms because it gives us a sense of identity or solidarity with the group or community or nation. These factors drive us over and over again to behave well—to help another, or contribute toward some common goal, even when it costs us something to do so.

Third, we perform selfless acts because we are emotional beings just as much, if not more, than we are rational beings. Our needs and desires are far more complex and varied than the simple desire to maximize pleasure and gain. Until recently, these findings have been largely ignored outside of psychology. But finally, these emotional motivations—our concern for what is right and fair, our desire to be seen favorably by our peers and so belong to a social group—are finally beginning to get their rightful due in other disciplines. Even economists, long in the habit of seeing human behavior as a series of supply and demand curves, or a string of *x*'s and *y*'s, are finally recognizing the power of emotional, social, and moral motivations, and beginning to challenge the status quo.

The promise of cooperation is not some silly utopian dream. It is grounded in some of the best work and most rigorous research in behavioral science. I will try to help us overcome our view of human nature—that all people are motivated by self-interest—and open your mind up to new possibilities. To do so may feel disturbing; one of our long-held assumptions is being challenged. But the evidence is on your side. It is time we all recognized this, and started acting on it.

Why Now?

I believe we are ready to break free of the selfishness myth and embrace human cooperation as the powerful and potentially positive force that it is. We as a society are in the midst of great disruptions in technology, business, ideology, and science. Ideas, trends, practices, and habits of mind tend to cohere over periods of time. When any relatively stable and coherent system—an economy, a country, or a community—suffers a shock, it leads to a new flexibility, a new openness to different ways of explaining our world and organizing our lives. This is the way we come to reexamine old practices, try new ones, and adapt to the changes happening around us. The economic collapse of 2008 has forced all of us to come to terms with the fallibility of economic and financial systems built on self-interest. This doesn't mean we're all about to become socialists (as some on the right might suggest). It simply means we should open our minds to the possibility that the Invisible Hand of the unregulated markets is a poor explanation of actual markets and actual human beings. It means that we should look to ways we can harness cooperation and collaboration to improve the systems we inhabit, rather than stubbornly cling to impoverished descriptions of those systems.

The world is changing at lightning speed. We are now at a period in our history when we need to learn how to rely on one another more than ever. But the seismic disruptions around us aren't all unhappy stories. Take what is unquestionably one of the greatest disruptions since the Industrial Revolution—the Internet. A few years ago, when I wrote my previous book, *The Wealth of Networks,* I spent more than five hundred pages trying to work out, in excruciating detail, whether and how the Internet is a fundamental, long-term change or simply a newer, faster vehicle for accessing, sharing, and disseminating the information we already had available. What I found was that the Internet has allowed social, nonmarket behavior to move from the periphery of the industrial economy to the very core of the global, networked information economy. Information and news, knowledge and culture, computer-mediated social and economic interactions form the foundation of everything in all aspects of our lives—from the pursuit of democracy and global justice, to the latest trends in business and media, to the best innovations in the most advanced economies. The Internet has revolutionized how we produce information and the knowledge foundations of our society.

The emergence of social production on the Internet has given us countless newer, cheaper, easier, and more rewarding platforms for collaboration than we have ever had before. On the web, people are engaging in voluntary acts of cooperation every day. We query strangers by searching for information on Google, and we get answers from people we don't know, will never know, for reasons we need not ask about, for free. We share our advice, and receive advice from others on everything from what movie to see, to where to buy a used bicycle, to how to cope with our child's illness. Whether it's via Wikipedia, a blog, or our Twitter feed, we volunteer our knowledge and expertise and expect nothing in return. The anonymity of the web makes us feel safe enough to join support groups for problems we would have otherwise suffered alone, find people who share interests we were embarrassed to indulge offline. We help one another expand our professional networks, even brainstorm with people in far-flung corners of the world to solve social problems or find cures for those diseases the drug companies have ignored. In other words, we find small ways and large to integrate voluntary, productive activities into our daily lives—and we use and build upon the contributions of others who do the same. Few simple metrics show this more clearly than the fact that almost all of the top twenty most trafficked and linked websites are either search engines like Google and Yahoo, social networking sites like Facebook, or sources of information or entertainment created by users like Wikipedia, YouTube, or Flickr. If you think of the top ten results on any given Google search, again you will find a large component of nonprofit, or personal, and in any event free and open sites providing the ultimate answer. And increasingly we see software developers, entrepreneurs, and civil society organizations experimenting with and building online systems of social, cooperative interaction—with amazing results.

The disruptions wrought by the Internet, of course, have rapidly accelerated the rate of globalization and scientific growth, all of which are forcing increasing numbers of businesses to examine how they can emphasize learning and innovation rather than mere efficiency. Only in slow-moving markets can efficiency alone keep you ahead. When you do not know where tomorrow's competitor will come from, and what they will know that you don't, you have to continuously learn and experiment. And this cannot be done through command and control, nor through incentive schemes (however well refined they may be), because creativity and insight are impossible to monitor and measure. More businesses are coming to see that they must instead harness the intrinsic motivations of their employees and managers—the moral, social, and emotional drivers of human behavior, rather than just the material. This not only makes employees happier and more productive, it also spreads the practice of cooperation throughout the entire industry by encouraging competitors to construct and employ their own cooperative systems and techniques.

I am optimistic in thinking that we are now ripe to take on the task of using human cooperation to its fullest potential—to make our businesses more profitable, our economy more efficient, our scientific breakthroughs more radical, and our society safer, happier, and more stable. The time is ripe not just in the world, but also in science. Today we have more evidence, more theory, and more legitimacy for the position that, while self-interest is of course part of the story of what drives human action, it is only part. A more complete story lies beneath the surface. Though we can see so many strides in human cooperation being made around us, there is even more evidence waiting to be uncovered.

What I hope this book will do is peel away those layers and help us reexamine our motivations and habits of mind. My hope is that by presenting the overwhelming evidence not just from the real world but also from the biological and social sciences, I can help us to overcome the power of the notion of universal selfishness and present new ideas and ways of thinking about how to design the systems in which we live: from the simplest business practices, to the most complex educational models, to the labyrinthine legal and technical arrangements governing everything from wireless communications to intellectual property controls on innovation and creativity.

For decades we have been designing systems tailored to harness selfish tendencies, without regard to potential negative effects on the enormous potential for cooperation that pervades society. We can do better. We can design systems—be they legal or technical; corporate or civic; administrative or commercial—that let our humanity find a fuller expression; systems that tap into a far greater promise and potential of human endeavor than we have generally allowed in the past.

Make no mistake. I will not be presenting you with a naïve, Pollyanna view of the world. The evidence I describe certainly includes robust, repeated findings that some people *will* act selfishly and in pursuit of their own self-interest some of the time. Moreover, no matter how selfless we may think ourselves, in certain situations almost all of us will look inwardly and say "of course I take material and practical payoffs into consideration." We know that punishments and rewards can help drive desired behavior; we also know they can undermine it, by suppressing our internal and social motivations to work together. We know that the same incentives, extrinsic or intrinsic, will not work for the same people at all times. People are different. Some are inclined to behave cooperatively, and so respond strongly to the various drivers of cooperation. Others do not. My aim is not to depict some fantasy world in which we pretend to be completely self-sacrificing creatures. It is merely to show that people, in general, will react cooperatively in certain situations and selfishly in others, and to help us figure out how to design systems that encourage, foster, and sustain cooperation to the greatest extent possible.

In the next few chapters I will look at the intellectual arc of work in various fields over the past fifty years in several core disciplines concerned with human action and motivation. We will look broadly, but also dive more deeply into the role of cooperation in social relations: the effects of empathy and solidarity, our drive to do what is right and fair, and our desire to conform to the normal. I will draw from such diverse fields as evolutionary biology, experimental economics, psychology, organizational sociology, and neuroscience. I'll also draw from the real world, with examples ranging from the band Radiohead's online pricing (or non-pricing) structure to the success of the Obama campaign and case studies of companies like Toyota and Google; from the harsh realities of a group of lobster fishermen to the strides being made by companies simultaneously pursuing social justice and profit. These offer useful lessons about how to induce and sustain human cooperation in a wide range of settings.

How we see ourselves plays a significant role in what we end up becoming. The selfish view of the self is not only unflattering, it is also a self-fulfilling prophecy. This book is at least partly about regaining a more balanced view of ourselves. I am not suggesting we are saints. Self-interest and cooperation aren't mutually exclusive; quite the contrary. Valuing independence, autonomy, capitalism, and individualism do not automatically make us egocentric, egotistic, heartless beings. Cooperation and profit can coexist. Embracing this duality, learning how to remake our society around it and harness it for individual, corporate, and societal goals, is not only possible, it is imperative. And its time has come.

• CHAPTER 2 •

Nature vs. Culture: The Evolution of Human Cooperation

For the whole sensible world is like a kind of book written by the finger of God . . . and each particular creature is somewhat like a figure . . . instituted by the divine will to manifest the invisible things of God's wisdom. . . . [H]e who is spiritual and can judge all things, while he considers outwardly the beauty of the work inwardly conceives how marvelous is the wisdom of the Creator.

—Hugh of St. Victor (d.1141), *De tribus diebus*

Man . . .
Who trusted God was love indeed
And love Creation's final law—

Tho' Nature, red in tooth and claw
With ravine, shrieked against his creed.

—ALFRED, LORD TENNYSON,
In Memoriam A. H. H.,
Canto LVI (1850)

Since the days of Augustine, scientists, scholars, and theologians alike have looked to the Book of Nature as a window into God's mind. For centuries it offered shelter for early modern science, a safe way to conduct their work without attracting the scorn or suspicion of the religious zealots. Human evolutionary biology over the past 150 years continued this tradition: debating core moral questions about ourselves in terms of natural sciences (in particular, evolutionary biology). Questions like: Are humans inherently selfish or altruistic? Is life fundamentally a competitive, zero-sum game, or an arena of cooperation? Are humans basically equal, or do we live in a naturally ordered hierarchy of superiors and inferiors?

The modern version of the debate goes back to the time of the Social Darwinists. Its political implications were never subtle. Herbert Spencer coined the term "survival of the fittest" to describe the idea that only the best-adapted species would survive and reproduce, and then used it to justify the harsh nineteenth-century version of laissez-faire industrial capitalism. Francis Galton used it as the basis for eugenics, the idea that the human race can and should improve itself through selective breeding. The idea was used most notoriously to justify extermination policies in Nazi Germany, but it also formed the foundation of "scientific" and "progressive" policies in many other countries, including the United States. Nowhere was this clearer than when the state of Virginia invoked a "science based" argument for sterilizing a mentally disabled woman in the 1920s. The U.S. Supreme Court upheld the policy in an opinion authored by the great progressive justice Oliver Wendell Holmes, Jr., who wrote: "The principle that sustains compulsory vaccination is broad enough to cover cutting the Fallopian tubes. Three generations of imbeciles are enough." That women belonged in the home, the natural low intelligence of the lower classes, the avarice of Jews, or the inferiority of African Americans—all these beliefs were justified using this scientific framework. No contemporary discussion of the biology of human morality can afford to forget this history. Nor can we afford to overlook the fact that our elevation of science as arbiter of truth in contemporary culture has so often caused us to abandon our moral judgment; that when "nature, red in tooth and claw," shrieks against our creed, we have often taken that to mean we should reexamine our creed, not our interpretation of nature.

If Social Darwinism anchored the "nature" side of the century-old nature/nurture debate, it was the work of American anthropologist Franz Boas that anchored the nurture side. His work looked to culture, instead of biology, to explain human differences across societies and races, spawning the birth of modern anthropology as a new scientific basis from which to attack the assumptions that certain races, religions, and cultures were superior to others. The first generation of this modern scientific nature/nurture debate continued for several decades, and it was only the horror of Nazism that put it to rest, at least for a time. Repulsed by the Nazi uses of eugenics and scientific racism, the scientific and academic community had by the 1950s finally settled the battle more or less completely on the culture side.

• • •

But it was not forever that the biologists would be banished from the debate over the mysteries of human nature and sociality. The vehicle that brought them back into the picture in the 1970s was the resurgence of interest in animal behavior. After all, no one really doubted that questions such as why starlings fly into the evening sky, or lemmings jump off cliffs, must be answered in evolutionary and genetic terms. That much was obvious. But how did the evolutionary logic that explained these behaviors translate to the human animal? No one was more influential in answering that question than entomologist E. O. Wilson, whose 1975 tome, *Sociobiology,* essentially relaunched the practice of applying evolutionary theory to human behavior and introduced the basic arguments that played out more popularly in Richard Dawkins's *The Selfish Gene* (and more recently in the extensive and popular work in evolutionary psychology). Genes,

Wilson argued, had human culture on a leash. Human behavior developed and could be explained by evolutionary forces, and evolutionary forces acted on genetics. Therefore, behaviors caused by genes that were favored by evolution survived and proliferated, and those that weren't, didn't. It wasn't long before Dawkins took this broad assertion and aimed it at the fundamental question of human nature that concerns us here: Are we "by nature" altruists or selfish? The title of his book is unambiguous as to his conclusion. Later, pioneers in evolutionary psychology* took this one step further; they attempted to explain human psychology as a series of evolutionary adaptations in the brain that code for specific human behaviors—ranging from a universal instinct for language, to gender differences in mating preferences and aggression.

In other words, Wilson's work punched Franz Boas's turn to culture in the teeth. But Boas's present-day followers were quick to fight back (in one infamous case almost literally, as activists assaulted and poured water on Wilson at a public lecture). As a result, a substantial amount of work in the 1980s focused on refuting and rejecting what was seen as Social Darwinism in a new guise. Critics of sociobiology, such as Stephen Jay Gould and Richard Lewontin, said that Social Darwinists used too naïve a version of adaptation (the idea that our genes adapt over time to meet the challenges of our environment), one that assumed too readily

* Such as Leda Cosmides, John Tooby, and Steven Pinker.

that everything is a product of adaptive selection, rather than an unintended by-product in a complex interaction. Historians of science focused on the moral abuses of Social Darwinism and cautioned that its revival would threaten recent gains in gender and race relations. In university campuses throughout the United States the debate was highly politicized, precisely for this reason. But the possibility that science could, after all, answer our most fundamental questions about human nature proved too seductive for scientists, scholars, and society to resist.

So today we continue to scrutinize the Book of Nature for clues as we try to understand ourselves. We still look to genetics, evolution, and the animal world for answers to questions about human morality, motivation, and action—particularly when it comes to the question of whether we are wired for cooperation. Take, for example, the opening words of a BBC online science report that I happened to read as I was writing a version of this chapter: "Ants are renowned for their ability to work together, and put the good of the community ahead of personal concerns. But new research suggests that their colonies are actually hotbeds of devious, selfish and corrupt behavior. . . ." Similar headlines can be found almost every week in one publication or another as we cling to neat and tidy evolutionary explanations about human nature. The surprising aspect of our turn to biology in the last decade, however, is that it tells a story about human nature and social behavior that is more sophisticated in its understanding of the interaction of culture and genetics, and, no less important, gives much greater play to cooperation. It rests on an assumption more reminiscent of Peter Kropotkin, who in a 1902 book, *Mutual Aid,* documented widespread cooperation in the animal kingdom, than it is of Herbert Spencer.

What Is the "Selfish" Gene?

The best way, perhaps, to see the arc of the debate over the last thirty years concerning the role of evolution and genetics in shaping human morality is to compare two very strong statements, made by two highly accomplished scientists, at the bookends of this period. The first is from Richard Dawkins's *The Selfish Gene,* in 1976. The second is from a review of the discipline of the evolution of cooperation, published thirty years later by Martin Nowak in *Science:*

> Be warned that if you wish, as I do, to build
> a society in which individuals cooperate
> generously and unselfishly towards a common
> good, you can expect little help from biological
> nature. Let us try to *teach* generosity and
> altruism, because we are born selfish. Let us
> understand what our own selfish genes are
> up to, because we may then at least have the
> chance to upset their designs, something which
> no other species has ever aspired to do.
>
> —DAWKINS, *The Selfish Gene,* p. 3

> Perhaps the most remarkable aspect of evolution is its ability to generate cooperation in a competitive world. Thus, we might add "natural cooperation" as a third fundamental principle of evolution beside mutation and natural selection.
>
> —MARTIN NOWAK, "Five rules for the evolution of cooperation," *Science* (2006)

What explains the stark difference between these two assertions? Has evolutionary biology changed so much in thirty years? Was Dawkins just overstating the implications of the science? Were the editors and peer reviewers of *Science* asleep at the wheel when they let Nowak publish his words? The answer is at least somewhat "yes" to the first two questions. Evolutionary biology has changed in the last decade and a half, and mainstream scientific thought has seen a resurgence of interest in, and a reinterpretation of, the study of cooperation. In particular, there are two central concepts about how the proclivity for cooperation has developed as a human trait. The first is a much-expanded understanding of how reciprocity factors into cooperation (beyond "you scratch my back and I'll scratch yours"). The second is a revival and reincorporation into mainstream evolutionary theory of the idea of group selection, the idea that some traits can evolve even when they are bad for the individual, as long as they increase the success of the group to which that individual belongs (which by the 1970s was considered a dead branch of evolutionary theory). Moreover, Dawkins often wrote in terms that lent his work to be somewhat misinterpreted. "Selfish gene" does not imply "selfish person." Under the right circumstances, in fact, nonselfish, cooperative human attitudes and behavior can be "selfish" if they increase that person's chance of surviving and reproducing. (However, when Dawkins then writes that biological nature will not help us to attain "a society in which individuals cooperate generously and unselfishly towards a common good," he does seem to say that selfish *behavior,* not just self-reproduction, is the result of the genes themselves. And that is a very different kind of statement.)

So the first thing we need to understand is that evolutionary biologists define the terms *selfish* and *altruistic* slightly differently from how we generally use them in everyday life. We might think of a selfish act as something like refusing to share that last piece of birthday cake, or accepting a multimillion-dollar executive pay package after laying off hundreds of workers and running your company into the ground. But in fact, a "selfish" act, as evolutionary biologists define it, is simply any behavior that maximizes one's chances of passing on one's genes to the next generation. According to that definition, even acts that we would in normal language call altruistic are "selfish" from the perspective of the selfish gene. Take this simplified example. Imagine that there were an altruism gene (we'll call it the AS gene) that predisposes people toward selfless behavior. Let's say that if you had the AS gene you would not think twice about the costs and benefits of giving to others, and you get an emotional high from seeing other people's suffering alleviated, so you would spend much of your life dedicated to the well-being of others, without a thought to yourself. Let's also say that you lived in a society in which when people observed these acts they trusted you more, gave you more opportunities, expressed more respect, and maybe also predicted that you would be a good parent. People who possessed this gene would therefore likely be considered a more desirable mate, and thus be more likely to pass on their AS genes to the next generation, who would possess the same advantages. In this example a gene is coding for a single behavior (if that were possible) that is clearly "altruistic" in the social sense, yet clearly "selfish" in the genetic, evolutionary sense. If we look at it this way, if we then try to "to build a society in which individuals cooperate generously and unselfishly towards a common good," we most certainly *can* "expect help from biological nature."

Current evolutionary science is beginning to offer us new, revealing insights into cooperation. It is helping us explain not only why acting cooperatively, or altruistically, increases our individual chances of passing along our genes, but also why groups can benefit from having strong cooperative practices and proclivities. In other words, it is beginning to help us explain why cooperative behaviors are passed down both culturally and genetically. In doing so, it offers us an interesting new angle for the nature-versus-nurture debate. Let's look at some of the work that sheds light on whether cooperative behaviors are genetically programmed, culturally taught, or both.

Drowning Siblings, Dueling Chimps, and Aesop's Fables: Reciprocity

The simplest theory for why cooperation could improve our individual genetic fitness is what W. D. Hamilton called inclusive fitness, later renamed kin selection. This is the idea that individuals will help others who are genetically related, because these are the people who will pass on at least part of their genes to subsequent generations. Take the easiest case. Let's say my three-year-old brother is lying in a shallow pool, facedown and about to die. Assuming I am relatively fit and will not dissolve in water, my risk of dying by pulling him out is very low, and since he shares half my genes, the benefit to my genes if I save him is high. So I am likely to save him without a second thought. But what if he is drowning in the ocean, and I'm a bad swimmer? The benefit to my genes if he lives is still the same, but now the risk is much higher. Or, worse yet, what if I know that, for whatever reason, he will never have children? Or what if he is adopted, and so we share no genes whatsoever? Or what if he is not my brother, but a second cousin? According to the kin selection theory, as the chances of him passing on my genes go up, so does the likelihood that I will risk my life to save him. J. B. S. Haldane put it in less than romantic terms: "I will jump into the river to save two brothers or eight cousins." This kind of altruistic act—saving my brother's life—is easiest to reconcile with what Dawkins would describe as strictly "selfish" at the genetic level, because it increases my likelihood of my genes surviving.

But how do we explain helping individuals to whom we are not genetically related? How can this possibly improve our genetic fitness? Here reciprocity (the notion that cooperative acts will ultimately be reciprocated) provides an explanation. The simplest version is direct reciprocity, an idea introduced by Robert Trivers, whose work inspired Dawkins, Steven Pinker, and many others. Basically this says that when we are in a situation in which another individual can directly reciprocate our help by helping us back, we both recognize that we will be better off by cooperating than not. This basic dynamic is not confined to humans; it has been demonstrated through numerous animal studies. One of the most gripping studies was described in Frans de Waal's fascinating book, *Chimpanzee Politics,* in which De Waal studied the shifting alliances among three male chimpanzees who were fighting for power: Yeroen, the oldest; Luit, the next in line; and Nikkie, the third in line. Nikkie, it turns out, was quite clever. He helped Luit dethrone Yeroen and then forged an alliance with Yeroen to overthrow Luit and become the dominant male (sounds a bit like a reality TV show, doesn't it?). The stakes were clearly high; rank in the chimpanzee hierarchy translates into opportunities for mating. Yeroen, Luit, and Nikkie were not brothers or cousins; they were competitors. And yet they found ways of sustained cooperation in ways that improved their own fitness.

Here's another example from the animal kingdom that sounds less like an episode of *Survivor* and more like one of Aesop's fables. Let's call this the fable of the Badger and the Coyote. In the National Elk Refuge, in Wyoming, a group of scientists observed that badgers and coyotes were collaborating to hunt ground squirrels. Coyotes, who are faster and have a larger range, would scout for squirrels, and once they spotted one, would signal to the badgers. The badgers, who are underground hunters and catch their prey by trapping them in dead-end tunnels, would then know where to burrow and lie in wait. The ground squirrels were now caught between a hammer and an anvil. If they escaped the badger by going aboveground, the coyote would catch them. If they evaded the coyote by ducking belowground, the badger would corner them. If you were a squirrel, witnessing this unlikely partnership between the two predators would be frightening. But if you were a scientist studying cooperation in nature, you would be hard-pressed to find a better example of different species working together. So how would scientists explain alliances like Yeroen and Nikkie's, or the coyote and the badger's? In both cases, animals whose instincts drove them to these cooperative behaviors, however randomly to begin with, were able to have more mating opportunities, or be more efficient hunters and therefore have a higher likelihood of living to create offspring, and then would pass on whatever genetic oddity led to the behavior to the next generation. This kind of direct reciprocity is really just another example of the simplest transaction in the universe: tit for tat. If you scratch my back—or help me overthrow the alpha male, or catch the squirrel—I'll scratch yours.

A Letter from Ben Franklin: Paying It Forward

So what about the huge range of cooperative or altruistic human acts for which we can't reasonably expect direct reciprocation? After all, in human societies, it's often far too complicated to keep track of who did what for whom, who paid back how much for what, and so on. Yet unlike other animals, we cooperate with and are generous to (in the sense of doing things for people without obvious gain) individuals who are no relation, as well as people we can't count on to directly reciprocate. We stop on the side of the road and lend strangers a hand in changing a flat tire. We give lost tourists directions in our city. We donate to charities that serve individuals we have never seen. We offer expertise online to people about whom we know nothing more than a screen name. These behaviors require a much more expanded view of reciprocity—what evolutionary biologists call indirect reciprocity (a term introduced by Richard Alexander twenty years ago, in a book called *The Biology of Moral Systems*). The best example of this I can think of dates back to 1784, with a letter written by Ben Franklin to a certain Benjamin Webb. Franklin, moved by a letter from the young man, sent him a sum of money. In the letter that enclosed the money, he wrote: "I do not pretend to *give* such a sum; I only *lend* it to you." This loan, however, is of an unusual kind. He continued: "When you meet with another honest man in similar distress, you must pay me by lending this sum to him." If this idea sounds familiar, it may because it was also the premise of a popular book by Catherine Ryan Hyde and later a movie starring Kevin Spacey and Helen Hunt, called *Pay It Forward.* It was the tale of a child who sets out to mend the world by doing a good deed for someone, and having them "pay it forward" (instead of paying it back) by doing something for three other unsuspecting strangers. They would in turn pay it forward to three others, and so forth, multiplying goodness throughout the world.

Of course, this is essentially a modern-day fairy tale, and a naïve one at that. But it does make a very important point. A system that hinges on indirect reciprocity (paying it forward) assumes that the giver of today is likely to be the receiver of tomorrow, and that eventually, gains will come full circle. It can still be "selfish" at the genetic level, in Dawkins's terms, because it assumes that today's donor gets, on average over a lifetime, at least as much as she gives. But what happens when someone breaks that circle by accepting the good deed and not paying it forward, or back, in return? As Ben Franklin already understood, the system only works until someone breaks the chain. He wrote of his loan to Webb: "I hope it may thus go through many hands before it meets with a knave that will stop its progress." So paying it forward can perhaps work for a while, if you have a population of people who are generous and honorable. But that is a population that is very easily invaded by a "knave" who will happily accept the generosity of others and never give in return. Knaves can easily invade a population of Franklins, unless the Franklins find some way of spotting them and keeping them out of their circle.

This is why most cultures have developed various symbols to help them determine who is a cooperator (a Franklin) and who is a defector (a knave). The primary one is reputation. Consciously or subconsciously, we assign reputations to people based on their prior actions. If it becomes known that I am a volunteer firefighter, I will develop a very different kind of reputation than if I get caught holding up a liquor store. And once I develop that reputation (whether it be as a local hero or an inept crook), people in my population need not know me to know whether I am trustworthy and can thus be expected to reciprocate good deeds; they only need know my reputation. Biologist David Haig put it beautifully: "For direct reciprocity you need a face; for indirect reciprocity, you need a name." But things can quickly get complicated with indirect reciprocity. Imagine that I see Alice hacking into Bob's computer and transferring money from Bob's account to Charlie's. That seems pretty unambiguously bad, doesn't it? But what if it turns out that Bob is a notoriously slippery character, who has stolen Charlie's money in a way that is completely untraceable and could never be proven in court? Perhaps I would start to see Alice differently. Perhaps I would start to see her as a Robin Hood–like heroine. Keeping track of who has done what to whom and interpreting what we see people doing to one another can get really tricky, and starts to require language, symbols, memory, and a capacity to form moral judgment. As two of today's most prominent mathematical biologists, Karl Sigmund and Martin Nowak, put it in the most immodest terms, "We argue that the intricate complexity of indirect reciprocity provided the selective mould for human language and human intelligence." When we look at kin selection, direct and indirect reciprocity, and other models of reciprocity in populations that meet one another relatively often (what evolutionary biologists call "populations with structure"), cooperation turns out to work quite well. David Krakauer of the Santa Fe Institute put it to me this way, in talking about the trajectory of evolutionary biology: "We used to think that cooperation was hard to sustain and very rare; now we think that it is much easier to sustain with a little bit of structure, and therefore more common."

Taking One for the Team: The Role of Group or Multilevel Selection

Reciprocity (direct and indirect) can explain a lot of selfless behavior (selfless in the sense that it helps others at a cost to the helper). But there are still stories that simply cannot be explained by reciprocity at the individual level. Let's look at another example from the animal kingdom. Instead of a reality show or a fairy tale, this one sounds like the plot of a science fiction movie. *Dicrocoelium dendriticum* is a parasite with an odd life cycle. It spends a generation reproducing, sexually, in the livers of cows or sheep. Its eggs are then excreted and eaten by land snails, in whose bodies these parasites reproduce asexually for two generations. The parasites then exit the snail in large groups encased in slime balls, which are eaten by ants. Most of the slime balls end up encased in cysts within the ant, where they stay, waiting for the next cycle. One or two, however, enter the ant's subesophageal ganglion, or nerve center, where they can actually manipulate the behavior of the ant, forcing it to spend its nights at the tops of grass blades instead of returning to the nest. This increases the likelihood that the ant, and along with it the rest of the parasite colony, will be eaten by a cow or sheep. The parasite that forces this behavior will die before it can reproduce, but it dies for a cause; now a new life cycle can begin for those parasites waiting patiently inside the ant's cysts. Those rogue parasites who go to the ant's brain in effect become the bacterial equivalent of suicide bombers, or kamikazes.

Let's look at this in terms of evolutionary fitness. The parasites with the genetic predisposition to alter the ant's behavior do not live to reproduce that gene. And the other parasites, whose genes predisposed them to hang back and let the other guy sacrifice his life, *did* pass on their genes.

So, how would such a behavioral pattern survive? No form of reciprocity can explain it, because the individual does not survive to be paid back, directly or indirectly, in time to reproduce.

The answer is that evolution can also favor behavior at the level of *groups,* whose members will sacrifice themselves for the good of others in the group, even if it works against their individual chance of survival. For about three decades, that answer, what we now call group or multilevel selection theory, was considered inadmissible by mainstream evolutionary biology. True, a more primitive form of group selection (the idea that species survived, not individuals) had been quite common in the 1950s and early 1960s. But this idea was subject to extensive critique, so that by the late 1960s group selection was largely banished from the mainstream. Selection happened at the level of the individual, biologists insisted, and really, as we already saw from Dawkins, at the level of the gene. But the work of some dogged researchers, most prominently David Sloan Wilson, kept the question alive, and finally, after four decades of argument, theoretical refinement, and new evidence (plus reinterpretation of some old evidence), group selection has come back to life. Today the theory is not only respectable, it's essential to explaining many of the cooperative behaviors we observe.

Take our self-sacrificing parasite, *Dicrocoelium dendriticum* (an example I borrowed from a book that Sloan Wilson coauthored with philosopher Elliott Sober). Imagine that a group of fifty of them enter a given ant's body encased in a slime ball. Clearly, within this group it is not enviable to be the carrier of a gene that makes you migrate to the ant's brain and direct it to live at the top of blades of grass, since that means you will die before reaching the cow's liver to reproduce. But when we look at the situation from the collective perspective of the group of parasites that carry that gene, things might be different. Imagine that within a group of 50 parasites, 10 carry this gene. So the odds for survival are 9 in 10 for those who carry the gene. Once the parasite has gone to the ant's brain and died, now only 9 of 49 carry the trait, rather than 10 of 50. These nine individuals, however, have a higher likelihood of getting to a cow's liver and of reproducing than any other nine individuals in any group that does not have self-sacrificers among them. The nine reproduce, their trait is further amplified in the asexual reproduction stage (where they cannot mix it and dilute it), and the next generation may bear several groups that include individuals each with this trait. In other words, each parasite with the mutation *individually* has lower fitness within the group, but collectively the groups that have self-sacrificers in them are fitter, and more likely to reproduce, than groups that do not.

Soldiers and Voters: Cooperation and Coevolution

The theory of group selection has been very productive in informing the groundbreaking work of anthropologists

Rob Boyd and Pete Richerson, who have spent the last quarter century applying evolutionary models to cultural differences and trying to explain the interaction between genes and culture. What they found is that in a sense, culture traits evolve just like any other—a cultural practice that improves the group's fitness will persist, and one that doesn't will become extinct. The difference is that cultural practices can change, be adopted, and be copied much more readily and rapidly than genetic traits. As a result, groups separated by small distances and short time periods can develop widely differing cultural practices, and some groups develop practices that dispose them to cooperate more effectively, or on larger scales, than others. Not surprisingly, societies or cultures that encourage these kinds of cooperative practices will be more successful and likely to survive, especially in a challenging environment, than those that do not.

The most obvious example is where one culture develops cultural practices that honor warriors (think of the repeated references to "honoring our troops" in American debates over the Iraq War). Imagine two groups. In one, the practice of serving in the army is culturally valued; in the other, it's not. In the first group, people are willing to risk their lives to fight for their group (or if not fit to fight, donate their special skills, such as weapon making or intelligence); in the second, they aren't. If these two groups go to war, the outcome is never in doubt. Just as the group of parasites with the self-sacrificing mutation is now more likely to pass its genes on to the next generation than the group without it, the society that wins the war is now more likely to pass on its cultural practices to the next generation than is the society that loses. This does not mean that the cultural trait is always adaptive, or positive. It may well spin out of control and lead to unnecessary conflicts. Cultures then need to find ways of channeling these cultural practices of group loyalty and commitment to less destructive forms. Nothing brought this home to me more clearly than a calm stroll down a street in the Hague: I came upon a street in which a row of bars had TVs blaring; raucous Dutch Orange fans were riling themselves up, loud, boisterous, immersed in a Euro 2008 soccer game against Italy . . . as the police silently stood in the background, across the square from them, making sure that the channeled aggressive nationalism did not spill over into actual violence.

Fortunately for us, as humans and moral beings, cultural adaptation does not require that those who do better vanquish those who do worse. Cultural adaptation, unlike genetic adaptation, can expand through imitation, not only through survival and displacement. When one group sees that another is able to win a war, or live more comfortably, or avoid hunger more effectively, et cetera, by cooperating, it can begin to borrow and imitate the aspects of that culture that seem to lead to the success. Now, it is entirely possible that cultures will misinterpret this success and adopt and enforce practices that have absolutely no adaptive value, but these errors play the same role as mutation does in genetic evolution and can sometimes have unexpected benefits (think of how dietary laws like kashruth or halal in Judaism and Islam spawned more hygienic food processing and preservation practices). The point here is that cultural traits evolve by a similar (if faster) mechanism to genetic ones.

But if maladaptive cultural traits get extinguished faster than genetic ones, haven't we just come up with an evolutionary story for human behavior that says culture does all the work? What room is left for genes in the explanation? Perhaps it is all nurture, not nature, after all.

The problem is that an "it's all nurture" approach does not do well in explaining all the evidence. Let's look at one example of a cooperative cultural practice that seems too recent on the evolutionary time line to have been effected by genetics yet may actually have a genetic component: voting. Voting has presented the selfish rational actor theory with a puzzle for decades. The basic quandary is this: The probability of one individual's vote affecting the outcome of an election is negligible. So almost any cost of voting (such as the time or bus fare it takes to get to the polling station) will be higher than the expected gain. Yet every year, hundreds of millions of people around the world do it, practicing a clear case of the triumph of a culture that values cooperation (at least when it comes to this particular practice). So what could genes possibly have to do with a behavior that is at most two hundred years old (and in most places was only generalized to the entire population in the twentieth century)? In a 2008 paper in the *American Political Science Review,* James Fowler, Laura Baker, and Christopher Dawes used a sample of four hundred identical and nonidentical twins in the Los Angeles area (all the sets of twins in the study were raised together, which meant that the results weren't compromised by the effects of early upbringing, and overall, they didn't differ in socioeconomic status or political affiliation). Their study showed that genetic makeup plays a large role not in whom we vote for, but in *whether or not we vote.* When the researchers looked at actual voter records in Los Angeles County they found that identical twins were more likely to show the same behavior—either vote or not vote (again, who they voted for was irrelevant)—than nonidentical sets of twins. In fact, after running various statistical analyses, Fowler and his collaborators found that slightly more than 50 percent of the concordance in behavior was due to genetics.

How could there possibly be a genetic basis for a social behavior that has only existed for most of the population for a century? A hundred years is a mere blip on the evolutionary radar; a gene for voting couldn't possibly have evolved in so short a time.

But imagine that there were a genetic personality trait called "conscientiousness," which described the degree to which people were inclined to inhibit selfish impulses, be deliberative in their decision making, and act in compliance with duty. Now let's bring Boyd and Richerson's theory back into the equation. Imagine that over the millennia, some cultures rewarded and valued conscientiousness, while others didn't. In these cultures, people who had a genetic predisposition to be conscientious would thrive, and because they would be considered more desirable mates they would reproduce, which would mean there would be more of them in the population over time. These cultures, in turn, would be able to sustain cooperation more effectively, because people would be driven to do the right thing even when they weren't directly monitored, punished, or rewarded.

Over time, then, both the cultural practices and the genetic predisposition would become more common. So in a culture where suddenly a new behavior, such as voting, were not only possible, but also culturally interpreted as the right thing to do, you would imagine that people who carried the "conscientiousness" trait would now start to exhibit it by voting (among other ways). And so, assuming this personality trait were heritable, then you would have a plausible story for how a very new human behavior could have a significant genetic component.

Studies have shown that personality traits ("conscientiousness" being one of the five most widely used traits in classifying personality*) are in fact partly heritable. A few years ago, Thomas Bouchard and Matt McGue published an extensive review of twin, adoption, and biological studies that looked at genetic influences on psychological and personality differences. They concluded that on average,

---

* Along with extraversion, neuroticism, agreeableness, and openness.

personality traits were between 42 and 57 percent heritable while shared environmental factors (such as the shared environment of the home) were not correlated with personality at all. (In a similar vein, there is significant work that suggests that other social attitudes, such as conservatism or religiosity, are heritable as well.)

The point here is not that there is a gene for voting, or religion, or for any other singular behavior, but that there are genes that *predispose* us to vote, go to church, fight in a war, et cetera, under the right cultural conditions. As Dawkins put it a few years ago, there is a "gene for developing the kind of brain that is predisposed to religion when exposed to a religious culture." The same can be said for cooperation and other prosocial behaviors.

As we begin to learn more about the biology of behavior, we can begin to get a better grasp on what role genes can play in interaction with culture. Take trust, for example. Having the ability to trust is a key component in cooperation, because, as we'll see in subsequent chapters, people are less likely to cooperate with people they don't trust. One of the most startling animal studies of the past few years looked at the effects of the brain chemical oxytocin on trust formation in voles. The researchers compared monogamous prairie and pine voles, who required much longer periods of male-female closeness before they would mate, and polygamous mountain and meadow voles, who mated much more readily. What they found was that the monogamous voles had higher-density oxytocin receptors in multiple areas of the brain than did the polygamous voles. In other words, more trusting partnerships occurred between animals whose brains had better oxytocin uptake. This insight led to a series of studies to see whether oxytocin had similar effects in humans. In one recent study subjects participated in a "trust game" that worked something like this: One subject received a sum of money from the experimenter and could choose how much of this sum to entrust to his partner. The experimenter would then triple the amount given to the partner, who could then decide to give back to the first subject as much or as little as he chose of this new sum. The game was set up so that the more the first trusted, the higher the *combined* payoff he would get. But since the second player couldn't be forced to share that gain, he stood to lose unless the first player could trust the second to reciprocate.

As we'll see in a later chapter, experimental economists have drawn all kinds of conclusions about human behavior from observing how people play such games, but the point here is that when some subjects were given oxytocin nasal spray, they were more likely to trust their partners with higher sums. In other words what we see is that personality traits—such as a disposition to trust—seem to have a biological component, which also suggests a genetic basis.

So the circle is closed. Behavior is affected by the physical brain, which in turn is affected by genes. Culture places selective pressure on individuals to conform to certain behaviors, and conforming to those practices may be harder or easier for different individuals, based on their genetic predispositions. Over time, individuals who possess genes that make them better able to conform their behavior to what counts as desirable in that culture, because "it comes naturally," will become more common, and groups that have such individuals and have applied their tendencies to productive forms of cooperation will in turn survive. This is what Boyd and Richerson called gene-culture coevolution.

• • •

What we are left with, then, is a developing story of the combined effect of genes, social dynamics, culture, and evolution. Of course the relative influence of each is not entirely understood; it is still somewhat of a "chicken and egg" debate. E. O. Wilson believed that culture was shaped by genes, not the other way around. Boyd and Richerson, however, argued that humans' behaviors, emotions, and beliefs are shaped by both their genes and their cultural practices; culture affects genes, and genes affect culture. To them, genes may have culture on a leash, but the dog is large and strong enough to pull that leash in very wide range. Whether the chicken (culture) comes before or after the egg (genes), the point is that people who are predisposed to cooperate will adopt those cultural practices that foster cooperation, and that evolution favors those cultures that possess cooperative practices over those that do not. The result is us.

So how do we account for the fact that selfishness, cruelty, and even evil still exist in the world? There are two distinct reasons. First, systems of reciprocity aren't perfect, so there is always some room for some individuals to get ahead by being selfish at the expense of those who cooperate. Similarly, systems based on group selection do not require unanimous compliance by all individuals within the group to work, which leaves room for a percentage of people in every population who are disposed to be selfish (it is precisely this fact—the diversity of human behavior and the possibility of better cooperative systems—that is the driving reason for this book).

The second reason is that *good* and *cooperative* are not always synonymous. In fact some of the most cruel and inhumane acts people have committed against one another have been committed by deeply "cooperative" people. In our culture, being "cooperative" is generally thought of as being "nice" or "generous." In a word, "good." This is a cultural adaptation; one of the ways in which we increase cooperation is by appealing to norms. But what counts as "good" in one society can take on wildly different forms, with wildly different moral implications, in another. Take rabid nationalism, or gang warfare. These can lead people to act in ways that are highly cooperative, in that individuals are sacrificing themselves on behalf of the goals of the group. But looked at from outside the group these actions can lead to terrible atrocities. Think about the suicide bomber, or the gang member who risks his life to kill an opposing gang member. In other words, our tendency to cooperate does not make us "good" universally. But it does make us responsive to cultural cues and allows us to feel solidarity, empathy, and trust with those around us. It helps to make us sensitive to moral judgments of right and wrong or fair and unfair *as defined by our particular culture.* Whether these tendencies will be harnessed for what we would call good or not depends largely on how we interpret behaviors we observe in others as judged by our own social and cultural beliefs.

What then, are we to make of the work on the evolution of cooperation? On one level these new findings are critical, because they give us a scientific framework for understanding how we came to be as we are. On another level, they can be only indirectly useful, because one can still only crudely predict how each of us, as genetically and culturally diverse as we are, might respond in various situations. So if we want to "build a society in which individuals cooperate generously and unselfishly towards a common good," we need to remember we are not building a society for genes, but for people, with unique psychology, behavioral responses, and cultural practices.

• • •

People continue to look at the natural sciences as though they were reading the Book of Nature. We continue to search in science for ways of understanding our origins and true nature, and how we ought to live. This is not surprising. Science is the most powerful tool we have in addressing the core questions about ourselves and our purpose.

We cannot, and need not, banish that ambition. But we also need to understand the limits of the biology of cooperation. If we want to understand this essential behavior, we need to pay attention to the human behavioral and social sciences, to history, and to technology, law, and business. I will use a bit of all of these approaches in understanding what makes human cooperation work. Each reveals certain parts of the picture. And each imposes some constraints on our desire to confirm existing beliefs and ratify existing practices. None do so perfectly, but we have no better. We might as well get on with it.

• CHAPTER 3 •

Stubborn Children, New York City Doormen, and Why Obesity Is Contagious: Psychological and Social Influences on Cooperation

Washington Square Park. I was sitting with my kids in the playground. A group of children were digging and playing with trucks in the sandbox. One of the kids was pulling a truck out of another's hands. The two children tugged the truck back and forth until the first child's mother intervened, explaining patiently that he should share his toy nicely with the others. He let go of the truck and was rewarded with a toothy grin from the other child and a warm, approving look from his mother. I watched, bemused. This was downtown New York City. The mother was well dressed, with the air of perhaps an investment banker or a lawyer. And yet her behavior was completely predictable and natural; she approached her child's "negotiations" over his truck in a way that would have been completely wrong had she done the same in her negotiations over her annual bonus, or her client's stake in a deal. There was something almost beautiful in the simple scene; it so elegantly captured how social context shapes our responses to others, how we are socialized and socialize others around us, and how we learn to segment our lives so efficiently that normal, well-adjusted adults can act differently in different settings—transcending crass self-interest in some, while maintaining the ability to look out for our own interests in others.

If the preceding chapter gives us a way of thinking about the evolutionary forces that underlie human cooperation—that is, why people and societies like ours survived and thrived where other, less cooperative societies might not have—this chapter looks at the psychological and social forces that determine how cooperatively we behave from one situation to another. Just think of the basic forces at work in the playground. First, and most obviously, there is material self-interest—that truck is *mine*! Work on moral development is no longer "in vogue" in the field of psychology, but classic works of Jean Piaget, and later Lawrence Kohlberg, and even somewhat more recent work agree that morality and conformity develop over a child's maturation process. Very young children, before ages three or four, can only see the world as it relates to themselves, a state Piaget called egocentrism. Over time they develop the ability to recognize and follow rules, but it isn't until they gain the ability to see the world from the perspective of others, or to understand the more abstract ideas about what is right and what is wrong, that they develop what we would call morality.

Though there is much disagreement within the psychological community about exactly *how* morality develops (and details such as what educational approaches best help kids develop their moral sensibility), there seems to be a good bit of agreement about the general trajectory. There is also a good deal of agreement that even once these abstract ideas about right and wrong do develop, they are highly sensitive to context. Whether it's sharing the truck, or, more ambiguously, sacrificing one's pay raise or bonus for the good of the company in a hard year, how we construe a situation triggers a whole set of emotional and cognitive responses that determine whether we cooperate.

These kinds of contextual cues and responses are much more relevant when it comes to designing cooperative human systems—be they technical platforms like Wikipedia or Kiva, organizational frameworks like a firm's management structure, or laws governing some shared resource—than the evolutionary ones I discussed in the last chapter. After all, what we most want to know in practice is how, exactly, to get people to respond to a situation in a cooperative rather than a self-interested manner. However, as we will see throughout this book, there is no single way to get at this answer; sometimes we'll rely on theory, other times on rigorous experiments, and other times simply on anecdotal evidence of real-life systems that have worked. At least at the moment, it doesn't seem that we have a better approach than this. Our economic models, which assume people to be driven entirely by self-interest, clearly do a very partial job. The psychological and sociological models are more nuanced, but less precise. And the case studies cannot always be transferable or generalized. So in order to understand cooperation fully, we will ultimately need a combination of them all.

• • •

Psychology probes a wide expanse of human behavior, and more so, many would argue, than does economics, or management science, or human evolutionary biology. For one thing, it acknowledges that people are driven by a great many (often subconscious) needs, goals, or values—not just self-interest. What's more, psychology acknowledges the fact that the needs, goals, and values that drive our behavior are not fixed or static—they change from one situation to another. And finally, psychology recognizes that not all people respond the same way to the same cues and contexts. It explores the important role that individual differences and personality play.

Let's look first at those basic drivers or motivations: our needs, goals, or values. The most widely known theory of the psychology of needs is Abraham Maslow's classic hierarchy, with basic physiological needs at the base of the pyramid, ascending to safety, to love/belonging, to esteem, and finally to self-actualization. Another well-known theory of needs (which we will encounter in later chapters) is Edward Deci and Richard Ryan's, which posits that humans have three basic needs: autonomy, competence, and relatedness.

Goals play a similar role in driving human behavior but are thought of as more active—that is, we more purposely strive to fulfill them. The third type of drivers is "values," which tend to be more socially oriented. Values, according to Shalom Schwartz's extensive work on the topic, include, in addition to things like benevolence, universalism, and tradition, other more self-interested motivators like hedonism, achievement, and stimulation. Since "needs" include "respect by others and morality," while "values" include "stimulation," you can see why, broadly speaking, I might treat values and needs interchangeably. For our purposes, these subtle differences aren't so important. What's important is that we as individuals are motivated by a range of forces, some more conscious than others—besides just material rewards. Some of the rewards that drive us may be material (for example, you need money to avoid hunger or pain), while others are more social (for example, to gain respect or avoid shame). What matters to us most at any given moment can vary widely from situation to situation.

Let me give you an example. I was walking back to my car, having dropped my kids off at their music school.

Parking in front of me was a mother and her child. She was trying to get her son, who looked about two and a half or three years old, into the car. "If you don't stop whining and get into the car by the time I count to five, that's five cents off your allowance!" she said. "Five, four, three . . . two, one. All right, that's five cents. Now, get in the car or it'll be another five cents . . ." The kid was still standing outside the car, stomping his feet in defiance. That little boy didn't care about his allowance. At this age, he couldn't really comprehend it, much less use it to regulate his behavior enough to comply with his mother's demands. Losing money from his allowance was too abstract a concept.

I live in Cambridge, Massachusetts, home to Harvard University and MIT; I bet, I thought, this mom is either an economist, or, more likely, is studying economics. Her behavior was a dramatic example of the risk we run when we try to govern our lives around a model that is too simplistic in its assumptions about how people will respond to various motivators. When we think of all forms of motivations as "incentives," we make grave errors. My needs and goals and values in one situation may be very different from my needs and goals and values in another. That mother, instead of threatening punishment, could have tapped into the child's need for approval and affection by saying something like "I love you. Do you think you can be a *really* big boy and climb into the car all by yourself?" I could easily write a mathematical model proving this tactic would have been more effective—all I need is to add a parameter that says "boy values maternal love." But that kind of model is simply a way of writing down what I have learned elsewhere. It doesn't really help us understand human motivation. That is why we must turn to psychology.

We See the World Through a Frame

This brings us to the second strength of psychology: its attention to situational framing. Framing, quite simply, refers to our interpretation of a situation, relationship, context, or event. Anytime we make a decision to act, we have to first interpret the situation we're in. Even economists have grudgingly admitted this; behavioral economics describes it as the framing effect. Amos Tversky and Daniel Kahneman, the fathers of behavioral economics, explain that people will make different decisions depending on how a situation is presented. For example, when making a bet, people will risk different amounts depending on whether the bet is described as risking a loss or aiming for a gain (behavioral economists have found that people display what is often called "loss aversion": they will reject bets framed as potential losses, but accept that same bet when it is framed as potential gains). Countless experiments have demonstrated equally powerful framing effects in a wide range of contexts.

While "framing" is popularly known today through these kinds of "irrationalities," the situation and its impact on what we want and what we can or should do is a long-standing component of social psychology. Sociologist Erving Goffman termed it frame analysis. It is relatively easy to see the effects situational framing can have on cooperation, even in very simple settings. One particularly simple example is called the Wall Street/Community Game experiment. Psychologist Lee Ross and his collaborators divided their subjects into two groups and had them each play a standard Prisoner's Dilemma game. In a game of Prisoner's Dilemma, two people are each made the following offer: If player A cooperates and player B refuses, player B gets, say, $10, and player A gets nothing. If player B cooperates and player A refuses, player A gets the $10, and B nothing. If neither player cooperates, both get, say, $2, and if both cooperate, then they each get $5. The two players have to make a decision without knowing what the other player will do. Clearly they are better off if they both cooperate than if they both refuse to cooperate, which in the game is called "defect." But because neither can trust the other to cooperate, and each will act so as to make himself the best-off he can without relying on the other, game theory predicts very clearly that each will defect so as to make sure he gets at least $2, instead of taking the risk of getting nothing by cooperating. Many experimental games have shown that, in fact, real human subjects in the lab cooperate much more than theory would predict.

What was special about Ross's version of the game was this: They told one group that they were playing a game called "the Community Game" and told the other they were playing "the Wall Street Game." The rules were identical; the monetary payoffs were identical; the only difference was the name of the game (in other words, the frame). As it turned out, people's willingness to cooperate differed dramatically depending on which version they were told they were playing; those told they were playing the Community Game cooperated about 70 percent of the time, while those told they were playing the Wall Street Game cooperated only about 33 percent of the time. Simply using such culturally loaded words as *community*, which implies cooperation, "share nicely," and *Wall Street*, which implies being aggressive and self-interested, was enough to elicit those very tendencies. These labels may have swayed the participants' behavior by influencing their predictions about what the other players would do. In other words, if you know the other person believes they are playing a "Community Game," you might expect them to cooperate; then you might be more willing to take the risk of cooperating as well (even assuming you aren't the kind of person who is out for yourself, you don't want to be taken for the sucker, either). Whether it had more to do with the cultural connotations, or how people believed others would respond to them, one thing was clear: The framing of the situation was enough to produce major variations in the levels of cooperation. These studies were done with groups ranging from college students in the United States to pilots in the Israeli air force, with the same results.

In another interesting twist, when Ross and his collaborators asked the subjects' teachers and commanders to predict, based on what they knew of their students personalities, who would cooperate and who would defect, it turned out that how the game was framed predicted people's actual behavior far better than their teachers and commanders did. Those who were often self-interested and out for themselves in the real world could be moved to cooperate simply by a reframing of the experiment as a "Community Game"—and vice versa.

So now we have the two major psychological pillars for understanding what makes us behave cooperatively or selfishly: the fact that we have a diverse set of needs and goals, and the importance of the situation to determining how they are triggered (or "activated," as some psychologists would call it).

The third major psychological variable we need to consider when trying to predict behavior is individual difference and variation—or, to put it simply, personality. Most of today's work in social or cognitive psychology is not interested in individual difference, but rather in "average" responses. In other words, it focuses on how *most* people respond to a given situation, and what counts as an abnormal response to that average. Personality studies focus instead on variation—how can we predict what *this* person, as opposed to an average normal person, would do in a given situation. As with needs, goals, and values, there are quite a few different approaches to defining personality. The criteria most widely used to map personality are what psychologists refer to as "the Big Five": openness, conscientiousness, extraversion, agreeableness, and neuroticism. It is easy to see how some of these would predict an inclination to cooperate. For example, a conscientious person would be likely to follow social rules or cues about the right or proper thing to do in a given situation (that's why I used it in the prior chapter to explain modern cooperative practices, such as voting), and an open person would likely be more trusting and therefore more willing to work with others. These are fairly intuitive. However, there has not been much work specifically about the effects of certain personality traits on cooperation, nor on exactly how individual disposition makes someone more or less responsive to cooperative cues and situations.

What we certainly do see in experiments is that people behave differently in identical situations: Some are cooperative while others are not. What has not been well studied is whether the long-held folk wisdom that some people are basically selfish and others are basically nice and reliable and cooperative in fact holds up scientifically. Intuitively, it seems like it should. But we also know that "good people sometimes do bad things." And as we'll see in chapter 4, at times people who are basically cooperative can act in ways that are quite destructive to others. So when it comes to designing systems that motivate people to cooperate in a way that serves the collective good, where does this leave us? Clearly we need to look beyond psychology and explore some of the biological and social factors that play a role in motivating people to do good.

In addition to the more conceptual and observational experimental work in psychology into needs/goals/values/framing and personality, in recent years there has been extensive work in the field of neuroscience, attempting to understand how certain behaviors are tied to brain function. This new work, which uses sophisticated brain imaging technology as well as chemical and electrical manipulation of the brain, is more biological in orientation and makes clearer statements of cause and effect—for example, "if you give subjects oxytocin nasal spray they behave more trustingly." As a result, it is receiving a lot of attention and newfound scientific respect. As was the case with the rise of economics among the social sciences, I think this at least in part has to do with our attraction to clean, determinate answers (compared to the more loosely defined theories that have emerged from other branches of psychology). And while certainly it is interesting and useful to know which discrete areas of the brain "light up" when we have a certain emotional response, there are two limitations to this approach as a basis for planning functioning human systems. The first is that in many cases, *several* parts of the brain respond in unison to a given trigger or cue. Just as single genes do not code for whole human behaviors, so too is it unlikely that a full, rich, human reaction in response to a situation will be fully reducible to the biological mechanism that physically produces it. The second is that most areas of the brain are involved in various, quite different responses. Together these suggest that while the idea that consciousness is physical, and that we can peg each action or emotion to a specific region of the brain is (a) seductive and (b) almost certainly true as a matter of biology, the flexibility in response and interaction of areas of the brain is such that translating these findings into an effective plan of action is highly uncertain. If we are studying physical social phenomena such as addiction, or violent disorders such as psychopathy, then yes, biology probably provides extremely important insights. But when we are focused on the normal functioning of normal human beings in a range of human contexts, the benefits are likely smaller.

Neuroscience can certainly help to show that there is a reward circuit triggered when we cooperate with one another, and that provides a discrete scientific basis for saying that at least some people really want to cooperate, if given the choice, because it feels good. Kevin McCabe and collaborators showed that people are indeed rewarded when they trust others; James Rilling and his collaborators showed that our brains light up differently when we are playing with another human being than with a computer. But human behavior is very complex, and a neuroscientific, biological understanding of the brain simply isn't enough to fully explain precise patterns of behavior, and precise responses to social cues and contexts. Just as with evolutionary theory, brain biology applied to most questions of morality and behavior provides little in the way of concrete answers about how to improve cooperation in actual, functioning social systems. As we will see when we look specifically at drivers of cooperative behavior, there is some very productive and insightful work in neuroscience and biology that complements experimental and observational work on the role of empathy, or morality in human behavior, but we are still very far from being able to pinpoint the biological brain functions associated with cooperation. (Okay, maybe we should try pumping oxytocin into the air the next time the U.N. Security Council meets . . . interesting things might happen. But then again, it would more likely end up more in the ventilation systems of car dealership salesrooms than international treaty negotiations.)

Social Capital, Reputation, and Social Contagion

It turns out you can't apply for a job as a New York City doorman. There are no want ads, no job postings, no online applications. As Peter Bearman showed a few years ago, becoming a doorman in New York "is both impossible and too easy." Doormen in New York are recruited through social networks, and only from social networks. Why? Because the risk of hiring a doorman who is untrustworthy and might jeopardize the safety and security of a building's residents is so great that a conventional market system simply isn't enough to guard against it. A social network, and a network built on reputation, can better ensure that the person who is chosen won't take advantage of the position of trust. At least, that's what New York City co-op and condo managements seem to think.

The idea that reputation and social networks are critical to economic activity was the basis of the book *Getting a Job* by Mark Granovetter, which introduced the idea of social capital almost forty years ago. The idea is simple. There are some things that make you more valuable as an employee, or a manager, that have to do solely with your social connections, rather than with your education, skill, or effort. In other words, having an address book full of people who know people is an economic asset.

Social capital is one of three major social dynamics that can improve cooperation. Of these, it is the one that is most clearly in line with the self-interest hypothesis, as social capital can enhance your own payoff over time—for example, by helping you get a job. Still, while social capital can yield financial rewards, as in the doorman example, its benefits go beyond mere material incentives. In some ways, social capital is entirely different from financial capital, because you can't always buy social capital with money.

Take this hypothetical example. Imagine that you are going for a job interview at a fancy law firm. You have one of two resources to choose from to help you land that job. The first is an envelope with fifty thousand dollars in it, to be used to pay the hiring partner at the law firm in exchange for the job. The second is an envelope with a letter of recommendation from your uncle, who went to law school with the hiring partner, then practiced with her as a prosecutor for a few years early in their careers, and continues to meet her two or three times a year at various social occasions. Which envelope is more likely to get you the job? At least at large, established U.S. law firms, the answer is clear. If you offer the hiring partner the fifty thousand, she will see it as "a bribe" and consider you unfit for a job at the firm. If you offer the letter of introduction, it will likely help, and may even be the deciding factor. The point is that for some kinds of interactions or exchanges in society, having a social relationship is more useful and valuable than having money. Sometimes money is not useful because paying for certain things is against the law—as in the buying of votes. Sometimes money is not useful because of social conventions or ethics rules—as in the example above. The bottom line is that social relations have their own currency. In the case of the law firm, it might work like this: Your uncle spends his social capital to get you the job, and by hiring you, that hiring partner will, in turn, accumulate social capital with your uncle and with you. This capital will itself be redeemable for resources—perhaps an invitation to a function, or news of a career opportunity—that are similarly exchanged for social capital, and so on.

While the kind of social capital in this example takes place between individuals, social capital can also be exchanged in less direct ways—most notably, through reputation. Most people understand that there are benefits to being seen as kind, generous, and trustworthy; in fact, in economic experiments, people behave more cooperatively when they know that their behavior will be visible to other participants in the experiment, because they anticipate that people will treat them better later if they are known to be someone who treated others well in the past. In online cooperative systems, reputation mechanisms, such as eBay's reputation system, which ranks sellers according to how satisfied buyers were with past transactions, are widespread and play an important role in keeping people honest. So in designing cooperative systems, we can't underestimate the importance of incorporating ways for people to both build and display reputations.

But reputation and social capital are by no means the only social dynamic at work in a cooperative system. Another critical factor is social learning. Much evidence suggests that our behavior in any situation is greatly influenced by the behavior we've observed in people around us, even if we're completely unaware of that influence. Consider a seemingly simple, individual decision, such as whether or not to eat that next cookie. We've all been there. It feels very much like an individual decision, right? We either exert willpower, or fail to. We either eat the extra helping, or we refrain. Sometimes we look at others and think, "I wish I had that metabolism." I know I do. How surprising, then, to wake up one morning in July 2007 and find that our collective struggles with our weight aren't a result of individually deciding to eat too many cookies; in fact we actually "catch" obesity from our friends, siblings, and spouses! Well, that's not exactly true, but close enough to be interesting. In a fascinating paper, Nicholas Christakis and James Fowler studied obesity among more than twelve thousand people who participated in the Framingham Heart Study between 1971 and 2003. When they looked at social ties between the participants, what they found was that obese people tended to have obese friends, siblings, and spouses. At first this doesn't seem all that surprising. One might expect that obese people would tend to stick together. People who have the same eating habits, such as siblings or spouses, would become obese together. But the timing of the weight gain in this study, however, showed that this was not all that was happening. Instead people were becoming obese *after* their friends and family members had become obese. In other words, obesity was spreading through social networks like a virus. In fact, the study showed that one's friend becoming obese increased one's risk of becoming obese by 57 percent; having a sibling become obese increased one's likelihood of following suit by 40 percent; and with spouses, the risk increased by 37 percent. In short, the people were "catching" the eating behaviors of those around them.

It shouldn't come as a surprise that if social learning influences how much and what we eat, it would also influence to what degree we cooperate. One interesting example is a study on tax compliance. After the 1986 Tax Reform Act was passed, most lawmakers assumed that compliance with the new tax code would increase as the result of penalties and the probability of an audit. This, however, turned out not to be the case. Instead the single factor that predicted people's compliance wasn't the size or threat of the penalties or the chance of getting caught, but *whom they spoke to* in the months leading up to the implementation of the reform. If they talked to people who said they were going to comply, they later reported that they too intended to comply; the reverse was also true. Similarly, in one of the few experiments of its kind, when the Minnesota Department of Revenue sent letters to certain taxpayers informing them of the fact that a majority of their fellow citizens pay their fair burden of taxes (voluntary compliance levels in the United States are high, by global terms, at more than 80 percent), they found that, with all other factors taken into account, the taxpayers who received the letters declared somewhat more income on their taxes (fewer deductions) than those who didn't. When the Australian tax authority ran a similar study, sending taxpayers letters informing them about the fact that most taxpayers strongly believed that overclaiming tax deductions was wrong, it was even more effective in keeping people honest. As we will see in the chapter on conformism, part of this has to do with the fact that we seem to want to conform to social conventions. But to do so we have to know what others are doing. We have to be able to observe and learn what the socially appropriate behavior is; when we do, we tend to follow it. In short, when the company we keep is cooperative, we too are more likely to be cooperative.

As we will see in the next chapter, there is one more

social dynamic that comes into play in terms of cooperation. That is our capacity to feel solidarity with others, to feel part of a community and to care enough about that community or group that we are willing to sacrifice our own well-being for the collective good. This partly has to do with some of the psychological motivators I mentioned at the beginning of the chapter—especially our basic human need for human connection. We seem to be wired to need others; as any awkward teenager will tell you, few punishments are as debilitating and painful as social isolation (this is why solitary confinement is the ultimate punishment, even within a prison). Responding to our need for human connection can be incredibly useful in designing cooperative systems; again, as we'll see in chapter 5, simply building in ways for people to establish and strengthen human connections can vastly increase participation. Later, when I talk about some of the strategies being used by various organizations to foster cooperation, we will see ample evidence that the most effective ones involve creating a sense of community and social connection. We are social beings no less than we are individuals. Systems that let us be both work better than systems that see us as just one or the other.

• • •

In the next few chapters, I will provide more detail on the social, psychological, and even biological evidence that we are far more cooperative creatures than we have long believed. While we know a great deal from these disciplines about what drives us, all these threads have not yet been combined into a theory for understanding human action. We know that we care about ourselves, but also about others. We know that we are capable of empathy and feel a natural connection with groups. We know that we follow social conventions, as well as possess the ability, at some level, to act in a way that we think is right and fair. We know that we also care a lot about the social dynamics of a situation—about what others do and think of us. And we know that these facts are true to different degrees for different people at different times. But we are nowhere near being able to predict how each person will act at a given point in time. What we are able to do, and what I hope to spell out in the next few chapters, is to begin to bring together what we know about people to answer the basic question: How can we build cooperative systems that protect us against our worst selves, without relying on either the fear of punishment, or the strategy of Leviathan, or purely on incentives and carrots, or money—the Invisible Hand? In other words, how can we systematically use the science of cooperation to improve our Penguin strategies? How can we harness and motivate one another to be our best selves?

• CHAPTER 4 •

I/You, Us/Them: Empathy and Group Identity in Human Cooperation

My family never got past the opening scene of *Empire of the Sun.* In that scene, the protagonist, a boy in Shanghai in World War II, loses his parents in the panicking crowds fleeing the town from the approaching Japanese troops. He turns around; there is nothing but a sea of adult legs and backs, none belonging to his parents. My son couldn't continue watching. The visceral, immediate sense of panic and pain that he shared with that little boy, his age, in Shanghai, was simply overpowering. I should have known. I also tear up, say, when I see Tevye saying farewell to Hodel at the train station in *Fiddler on the Roof,* as she follows her love to Siberia. Just writing about this makes my stomach clench. If you're a fan of the movie, maybe just reading it did the same to you. We're like that—we experience others' pain as though it were our own.

As human beings, we care about others. We care for our children and our parents. We care about our siblings and our friends and our colleagues. Most of us even care, to some degree, for those we have never met—even for people who aren't real, such as the characters in a movie. Who among us has not encountered an image of a child amid the wreckage of a war zone, or a hurricane-flattened home, or a parent weeping halfway around the world over a child failing as a result of disease or hunger, and felt compassion and empathy for that person, at least a little? It's not just that we imagine what their condition must be like; we also have an instinctive, emotional response. We *feel* for them.

Empathy is an extensively studied phenomenon. Some of the earliest work, done in the 1980s, focused on the most basic phenomenon—babies who begin to cry when they hear other babies cry. Much of that early work explored how our capacity to empathize develops both cognitively and emotionally. Martin Hoffman and Nancy Eisenberg, psychologists who have done some of the most extensive work on empathy, define it as a combination of cognitive and affective responses that work in tandem to identify *and then replicate* the emotional state of the other person. It is different from sympathy, or simply feeling sorry for someone else; it is our capacity to actually mirror and experience their feelings.

New advances in neuroscience have helped to identify some of the biological foundations for empathy. In one particularly powerful study, Tania Singer and her collaborators gave each member of a loving couple small electric shocks at alternating intervals and looked at their brain responses. When women were shocked, their brains were activated in three distinct areas. One region of the brain processed the physical pain while the other two were involved in the emotional response to the pain they were receiving. The scientists had predicted this. The surprising finding was that when their partners were shocked, these women showed the exact *same* activation, in exactly the same emotional areas, as when they themselves had been shocked. Only the part of the brain involved in the physical experience was not activated. In other words, these women literally felt the emotional pain of their partners. This was a particularly strong example of the phenomenon of neuron mirroring, originally identified by neurophysiologist Giacomo Rizzolatti. He found that when we observe other people doing something, the neurons in our brains fire in almost the exact same patterns as when we are doing those same acts ourselves.

As it turns out, our brains mirror not only pain or motor movements, but pure emotions as well. When Rizzolatti and his collaborators showed subjects videos in which other people were facially expressing disgust, the same neurons in their brains fired that were activated when they themselves were exposed to disgusting smells. In other words, there seems to be a biological basis for what my son was experiencing when we watched *Empire of the Sun:* that both cognitively and emotionally, we really can "feel" what others are feeling.

The interesting thing about empathy, and what makes it so distinct from "solidarity," attachment to a group (a topic to which I will turn in a few pages), is that it shows we care about human beings simply *as* human beings, regardless of who they are (or whether they even exist). When James Rilling and his collaborators scanned people's brains as they were playing a game of Prisoner's Dilemma with other subjects whom they had never met (and would never meet again), he found not only that people responded emotionally to these strangers, but that different parts of the brain lit up depending on whether a subject was cooperating with a computer or with another person.

Clearly empathy plays an important role in all manner of social behavior, including cooperation. The effects of humanization—of seeing the other person as a fellow human being—on cooperation can be observed not only in brain scans, or by asking people what they feel, but through the laboratory as well. Experiments have been conducted to see to what extent humanization motivates people to cooperate with others even when it is actually costly to them. One such classic study was done by economists Iris Bohnet and Bruno Frey. They recruited students who had never met one another and divided them into two groups. They gave the students in Group A ten dollars and told them that they could choose to take any amount of it home for themselves, and could put any portion of it into a sealed envelope, marked with a number corresponding to one of the students in Group B. After everyone put their sealed envelopes in a box, the Group B students would each take the envelope marked with their number. No one would know who gave what, or who received what from whom. So what do you think the students in Group A did? Obviously a selfish person would give nothing, since there was absolutely no way they could be penalized. Yet only 28 percent of Group A students in fact gave nothing; on average, the students gave about one quarter of the total amount. Next, Bohnet and Frey added a twist. They asked each student in Group B to stand up, so that now the Group A students had a visual image of the person who would be opening the envelope. Keep in mind they didn't speak to these people, didn't know their name, and would probably never see them again. No one, not even the experimenters, would know *which* Group A students had given which envelope—that is, who had been generous, and who had not.

How much did the students share now? Again, if we are to believe the traditional economic model that assumes everyone is out only for himself or herself, the students would still be expected to give zero, as there were still no consequences for looking out for their own self-interest. Yet it turned out that simply seeing the other person "in the flesh" was enough to drive the percentage of Group A subjects who gave nothing down from 28 percent to 11 percent and bring the average donation up from 25 percent to 35 percent.

Next, the experimenters humanized the transaction even more by telling the students in Group A personal information about the Group B students: his or her major, hobbies, et cetera. Amazingly, this was enough to raise the average amount given away to 50 percent; *not a single Group A student gave nothing*. How do we explain this incredible increase in generosity? Remember, the giving was still anonymous, so fear of punishment or retribution couldn't have factored in. The critical factor could only have been the human one; the more the students knew about one another, the more they could imagine themselves in the shoes of the other—that is, the more empathetic they were. This, in turn, translated into greater generosity toward one another.

Perhaps no other psychologist has done more work on connecting empathy to altruism than Daniel Batson. The prevailing belief in psychology previously was that all altruism was "really" selfish—people were generous to others only to feel good about themselves and alleviate their own distress at seeing suffering in others. Through a series of experiments, Batson showed that this wasn't the case; people helped others even when it would have been easy for them to avoid this distress at others' suffering simply by leaving the situation, or to simply stop thinking about the other person. Moreover, Batson showed that people were more altruistic, and more empathetic, when he told them to try to imagine what the other person was feeling (which suggests "pure" empathy, not just a desire to feel good about ourselves or alleviate our own suffering).

What his work, and others' along these lines, did in many senses is collapse the distinction between "altruism" and "selfishness." Whether we have an internal or "selfish" motivation to help others, or whether it's empathy that drives us, the results are the same—both in our behavior, and, interestingly, in our brains. If, as we saw in chapters 2 and 3, we are beings whose brains reward us with a shot of dopamine and oxytocin when we help someone, does that make us altruists or selfish? The answer, as far as anyone building systems for human cooperation is concerned, is "who cares?" Whether or not we are in essence just going after that next shot of dopamine by being generous and helpful to others is simply irrelevant. Knowing, on the other hand, that we are indeed the kind of being that is physiologically and psychologically wired to have these feelings, to get a shot of pleasure from helping others, is really all that matters.

So what does the fact that we're wired to empathize with and feel good about helping others mean when it comes to designing human systems? How can we, practically speaking, harness this tendency to encourage generosity and cooperation in the real world? One obvious way is by humanizing the people who need our help, as Bohnet and Frey did in the experimental setting.

One organization that has successfully fostered greater levels of benevolence and giving by humanizing its participants is the microlending site Kiva.org. Kiva, which is based on the model of borrowing circles (which have long been a common practice in less developed economies), connects people in poor countries who need to borrow small amounts of money to start or support their business to potential lenders in wealthier countries. Now, just like in many of the experimental situations we've discussed, there are no material incentives for the lender here (the interest on the loans is negligible), and given that the loans are anonymous to everyone but the recipients, who are strangers living on another continent, the social rewards are negligible as well. So how does Kiva motivate people to lend to people in remote places whom they will never meet? How does it help them decide between prospective borrowers? How does it foster trust in the process and empathy for the recipients? Just as Bohnet and Frey did with their students, Kiva does this by giving more information to help humanize the recipients. The site is designed so that those looking to borrow money post a photograph of themselves, a short blurb about who they are, and what they plan to do with the money they would like to borrow. Prospective lenders can then search these profiles and choose to whom—and how much—they wish to loan. Even though these transactions are often conducted across continents and oceans (unlike borrowing circles), even this much personal information fosters enough empathy, and enough of a human bond, to generate impressive levels of participation.

But we need not look to such specific examples to

recognize the power of the human bond. If we look at our day-to-day lives, we see that we spend an enormous amount of time, energy, and money on humanizing the people with whom we work and transact. Why do organizations spend money on company picnics and holiday parties? Why do businesspeople get on an airplane and fly thousands of miles for lunch or a dinner with a client (especially with all the sophisticated telecommunications available)? Because, however underestimated empathy might be in traditional economics, successful businesses know that these kinds of face-to-face interactions are needed to build trusting, co-operative, and profitable relationships. Those face-to-face interactions, as we saw in the lab, foster those feelings that motivate us to working cooperatively with one another in a way that ends up being mutually beneficial to both parties. And we *want* to feel these feelings—that's why we build our lives very much in order to allow us to expand the network of people with whom we share experiences and common bonds.

Stand by Me: Solidarity and Social Identity in Cooperation

If empathy is what makes us identify with, and sacrifice our interests for, other human beings, then solidarity, or group identity, is what motivates us to identify with and sacrifice our interests for those of the group to which we belong.

Examples of this abound in our society. We see it in team sports; a batter in baseball sacrifices his chance to get on base so that he can advance or score a team member already on base. We see it in the very existence of the military, a system that depends on hundreds of thousands of young men and women making the ultimate sacrifice (that of their lives) in the name of their country. Our desire to align ourselves in this way may explain why one of the most powerful phenomena of the past two centuries has been the rise of the nation-state, which has superseded the clan, tribe, or village as the primary marker of modern identity. As Boyd and Richerson's influential theory of gene-culture coevolution, which we discussed in chapter 2, shows, our desire to cooperate with and contribute to the collective good of a group (whose symbols, be it the team jersey or the national flag, we adopt as our own) plays a central role in human evolution.

At the experimental level, my collaborators Dave Rand and Anna Dreber and I have shown that solidarity alone is enough to sustain cooperation in public goods games. We tested subjects by first having them play a dictator game, a version of the model that Bohnet and Frey used, without any information. In this case, players were identified as belonging to one of two groups—Democrat or Republican. After a few rounds of the game, we then sorted the subjects again, this time into three groups: those who were simply not generous—who gave nothing to either type of recipient; those who were generous and unbiased—gave equally to people who had the same or the other political identity; and those who were generous and biased—who gave, but gave more to the people who shared their political identity than to the ones who supported the other party. We then sorted the subjects into groups with others of the same type, gave them each a bit of money, told them that they were in a group of Democrats or Republicans, and had them play a public goods game.

A public goods game is like Prisoner's Dilemma, but it is run among several players, not just two. The participants can contribute to a common pool, which we then multiply and divide equally among the players. The more everyone contributes, the larger the pot we have to multiply, and the more they all get as a group. But each one individually is best off by having the others contribute while he keeps all the money for himself. Since we share the common pool equally, he gets a share of the "public good" to which they all contributed, and also keeps all his original endowment. That's why the theoretical prediction is that no one will contribute anything, just as in Prisoner's Dilemma the prediction is that no one will cooperate. In actual experiments, however, public goods games usually start at reasonably high levels of cooperation, with about half or so of the participants giving. But when these games are played repeatedly, cooperators in the first round see that some of the players are unfairly taking advantage of them (not contributing but sharing in the profits), and they stop contributing themselves.

As we will see in later chapters, there are other interventions that are well known to change this. But what we found about solidarity in our experiment was that even though we ran the experiments for forty rounds, much longer than it usually takes for this decline to set in, the generous but biased players who were grouped with like players (Democrat with Democrat, etc.) continued to sustain their cooperation without significant change throughout the period, while cooperation both among the ungenerous and the generous but unbiased declined over time at the rate of most other experiments. In short, knowing that they were in a group of others like them was enough to make those players respond to that emotional trigger and sustain cooperation even without communication, norms, punishment, reward, or any of the other triggers that we will look at in the coming chapters, and which are usually thought to effect cooperation.

Of course, unlike empathy, solidarity usually results in not only an "us," but often a "them," too. And we treat "us" and "them" very differently. Who can forget the amphitheater scene from *Monty Python's Life of Brian,* when Brian learns about the People's Front of Judea, the Judean People's Front, the Judean Popular People's Front, the Popular Front for Judea, until someone asks "Whatever happened to the Popular Front?" John Cleese looks around and they all shout, "He's over there . . . *Splitter!!!*" On a more serious note, the point is that, as psychologists Elizabeth Phelps and Mahzarin Banaji showed, we very quickly and subconsciously sort people into groups, “like us” and “not like us.” And this too seems to be biologically wired. When Phelps and Banaji scanned subjects’ brains as they were shown photos of white and African American faces, they found that, as they predicted, given America’s troubled race relations, for both white and African American subjects, areas of the brain associated with fear lit up when they were shown pictures of the other race. As economists Sam Bowles and Herb Gintis emphasized when they explored solidarity as a mode of governance, the potential dark side of our seemingly innate desire for group solidarity makes this aspect of human cooperation highly volatile.

More than a hundred million people have been killed in the last century in the name of solidarity: from the Holocaust, through Stalin’s repression, to the Rwandan genocide, to the lower-level but more persistent horrors of strife in Northern Ireland or the ongoing tribal warfare in the Middle East. At the same time, it would be a mistake to think that group identity and behavior is necessarily all about hatred of the other. Group dynamics at the national level plays an enormously powerful role in justifying and sustaining substantial levels of giving and redistribution of wealth (such as welfare and charity) as well as substantial contributions to the collective well-being (such as volunteering in inner-city schools or in the military). Even the countries that are the most generous in their foreign aid are vastly more generous to their own poor residents than to those who are more remote, even though those remote others may have far greater need. Solidarity is, in other words, a powerful force for both evil and good, whether we like it or not. As we build systems, we want to understand it so as to better harness it. As we encounter its dark side, we want to understand it to disrupt it. Either way, we need to account for it.

One of the features of solidarity that allows us to harness it to changing needs over time is the fact that we are capable of identifying with more than one group at once, and even of changing our affiliation group as conditions change. For example, during the 2008 presidential primary election, a study I did with my collaborators Dave Rand, Thomas Pfeiffer, and Anna Dreber from the Program for Evolutionary Dynamics at Harvard looked not only at the effect of shared political affiliation on cooperation, but also at the degree to which group affiliation can shift as challenges to the group change. When we recruited two groups of Democratic voters—supporters of Hillary Clinton and Barack Obama—and had them play a dictator's game, the subjects, as expected, gave more to those players who supported their candidate (even though it was made clear that the money wasn't going toward anything that directly or indirectly had to do with the campaign). When we conducted the experiment again, just before the Democratic convention (by which time it had become clear that the two primary candidates' goals and values were basically the same and that there was no practical benefit to having two distinct groups within the Democratic Party), the results still didn't change. Both groups were still treating each other as "other." But, it turned out, we are creatures who are very susceptible to symbolic performances. When we ran the experiment again immediately after the convention, the results differed dramatically. The members of the two groups no longer treated each other as "them" people, but as "us." Some, in particular young men, suddenly became hypergenerous to supporters of the other candidate, even when just days before they had given those players the short shrift. It seems that after witnessing the big public show of unity at the Democratic convention, Democrats' sense of identity shifted; now it wasn't tied to Hillary Clinton or Barack Obama, it was tied to the party itself and to bonding against the new "other"—the Republican Party.

Praying on Street Corners

One of the most interesting efforts to harness cooperation through humanization, empathy, and shared identity is in the rise of community policing. In the 1970s American cities seemed to be in universal decline, with crime rates inexorably rising. So by the 1980s, several movements had developed to reform policing. One of the most successful was what's called community policing, a set of initiatives by which residents and police officers work together to reduce crime in urban areas. Unlikely as it sounds, community policing turned out to be incredibly effective in overcoming cultural barriers and long-standing grudges, and in bringing members of a diverse and often at-odds community together for the collective good. No example illustrates this better than the story of Chicago's notorious West Side.

One of the highest-crime areas in the country, Chicago's West Side had long been plagued by terrible relations between its police, who had a reputation for being uncaring, untrustworthy, and even racist, and its residents. In the 1980s, following criminology studies of the area, the police department began to experiment with a very different approach. As political scientist Archon Fung later chronicled, this new approach emerged from the realization that the two groups needed each other: The residents had unique knowledge about the problems that were contributing to the high crime rates in the community (for example, the location of the newest crackhouse, or a poorly lit park favored by muggers), while the police had the resources to solve them. But in order to work together to reduce crime and make the neighborhood a better place to live, the residents and officers would first have to find some way to relate to each other as human beings—and that's just what they did.

Here's how. First, the Chicago police took some beat officers, dubbed "community specialists," off rapid-response (911) duty, giving them time to walk rather than drive their beat, which afforded more chances for face-to-face communication with residents. Then the community specialists began to hold monthly meetings with community members, in which they could share information about what was going on in the community. Once the residents overcame their initial distrust, those meetings turned into even larger, more open forums. Between the face-to-face communication created by the beat officers and the familiarity built through the monthly meetings, the police soon stopped being the "other" to the community, which freed both groups (just like the united Clinton and Obama supporters in my experiment) to focus on their common challenge—the criminals threatening their streets.

Less important in the overall twenty-five years of the program, but uniquely interesting for our discussion of humanization and solidarity, was one particular community policing initiative, studied extensively by criminal law scholar Tracey Meares, and which raised quite a few eyebrows: the prayer vigil initiative on Chicago's West Side. In 1997, Harrison district commander Claudell Ervin invited the residents of his district to a series of prayer vigils held on a number of drug- and crime-prone street corners. Now, in bringing the community and the police together in this way, Ervin faced two major obstacles. One was the continued general skepticism and distrust, born of years of strained race relations, of the police within the African American community. But Commander Ervin came up with a clever way to overcome this. Since most residents in the district belonged to churches, he called pastors in the area and asked them to help organize the vigil, and to call upon their congregations to attend. This worked surprisingly well. As one of the leaders involved said in an interview, "I don't believe it was really organized by the police. I believe it was organized by a Christian who happened to be a policeman. And that's a world of difference." The second problem was the insularity of the individual congregations themselves—they were very distinct, close-knit communities that tended to be wary of outsiders. So to help them band together and forge one common group identity (rather than remain as dueling factions), Ervin set up the meeting with the pastors at a neutral location—the police station—instead of favoring one church over another. Again, it was a relatively small symbolic gesture, but it was amazingly effective.

Now, standing on a crime-ridden street corner is a risky enterprise no matter how you slice it. And in this community it was perhaps even more risky to be seen collaborating with the police. In other words, people had a lot of reasons not to participate, to applaud the effort from the safety of their own homes rather than risk their own well-being out on the street. Yet hundreds of residents of some of Chicago's highest-crime neighborhoods—people of various denominations and backgrounds—put the good of the community ahead of their own self-interest and turned out in droves. The result? In the months after the vigil, both church leaders and police officers who were interviewed said the experience had not only created a strong bond between the police and members of the community, but it had also accomplished its wider goal, which was to increase participation in the community policing program. What is fascinating from a design perspective is that Ervin successfully used the church to break down the community-police barrier, and then used the community-police intersection to break down the interdenominational barriers. This is exactly the kind of steering that would be necessary, and tricky, in harnessing solidarity to positive social ends. It is difficult, but, as it turns out, also possible.

So, is community policing effective? This is a continuing puzzle. Because community policing was introduced at a very high-crime period and because throughout the United States crime has since declined across the board, both in communities that adopted these collaborative approaches and those that adopted more hierarchical managerial approaches (such as the CompStat system, pioneered by New York City police commissioner William Bratton under Mayor Rudy Giuliani), its overall effectiveness is difficult to quantify. What is clear, though, is that both clearly improve on the strictly punitive approaches, typified by "three strikes you're out" policies, and both are associated with declines in crime. What is also clear is that community policing has been enormously popular with both police departments and the communities they serve; more than two hundred thousand police officers in some of the highest-crime areas in the country are involved in community policing efforts, and throughout the country people in the communities where it exists report much better relations with the police and less fear of crime. So while it's difficult to determine its absolute impact on crime rates, community policing gives us a rich example of how cooperation can be fostered within a social and public institution by introducing humanization, empathy, and solidarity. It also offers an example of how a cooperative system can be as effective, if not more so, than purely punitive and hierarchical solutions while at the same time garnering greater overall acceptance and goodwill in the communities in which it is implemented.

• CHAPTER 5 •

Why Don't We Sit Down and Talk About It?

Economists like to think of talk as being "cheap." What they mean is that when people with conflicting interests talk to one another, if they don't bind themselves to a contract where promises are made and money changes hands, then letting them communicate amounts to a whole lot of nothing. What people say to one another about their motivations and intentions, economists claim, is more or less worthless information—it doesn't tell us anything about how people would actually act if something real were at stake. Because talk is supposedly "cheap" in this sense—it doesn't commit you to anything—a whole strand of work in economics has developed to place people in situations where they have to actually act, rather than talk: buy or not buy, invest or not invest, as this is assumed to be the only way to reveal their true preferences. That's what prediction markets, which have become so popular in the run-up to elections, for example, do; prediction markets are speculative markets—betting exchanges—created to make predictions. They are just one example of a field called "mechanism design," built on the assumption that it is not what people say, but what they do, that counts. It has been so influential that it won its inventors the Nobel Prize in economics in 2007. While much of the work being done in the field is of course laudable, and is sound in theory, talk in both experiments and the real world is in fact not at all "cheap" and is very far from meaningless. We spend an enormous amount of our day-to-day lives communicating and interacting with other people. These interactions play a much larger role in affecting how we behave than the standard "rational actor" model would have us believe.

In the previous chapter we talked about how empathy, humanization, and solidarity can help to foster cooperation. Well, as it turns out, these forces become even more powerful, and easier to sustain, when you add in the element that is the focus of this chapter: communication. While "talk is cheap" might be a catchy phrase (and of course there are times when it's accurate), when it comes to the science of cooperation, communication turns out to matter a great deal, both in controlled settings and in the real world. The evidence for this proposition is overwhelming. Most famously, a mid-1990s study by David Sally found that in more than one hundred social dilemma experiments, conducted with thousands of subjects over decades, the findings were consistent: Levels of cooperation rose by 45 percent when players were simply allowed to communicate face-to-face. No money needed to change hands, no commitments needed to be made—face-to-face interaction alone was enough to nearly double levels of cooperation. At about the same time, Elinor Ostrom and her collaborators documented their own experiments in which strangers thrown together were given the opportunity to talk to one another. When they did so, they readily made promises and commitments to one another, set norms to govern their behavior, and engaged one another as human beings. But their talk wasn't "cheap" or meaningless—they overwhelmingly followed through on their commitments, even when there was no way to enforce these commitments in the experiment.

The same thing happens in the real world. Take the poster child for online cooperation: Wikipedia. On its face, Wikipedia seems like the most impersonal platform one could imagine. After all, its contributors seem to be tens of thousands of strangers flung all over the world. It is mediated by a computer. Except among its most engaged contributors, it's basically anonymous, save for screen names. It is certainly disembodied. So it seems counterintuitive that its members would feel any sense of community, solidarity, or allegiance with one another. But they do, in large part because Wikipedia has so many channels for communication.

One such channel is the "discussion" pages accompanying every article. This is a completely open forum where Wikipedia authors and readers can point out errors or problems with another's edits or contributions to an article, ask questions, or simply create a human connection. Debates on the Wikipedia discussion pages often become just as heated and lively as any face-to-face debate. For example, look at the following argument, which I drew in mid-July 2008 from the discussion page for an article on George W. Bush. Over the years the entry has been edited and reedited many times, sparking a good deal of controversy. Here is a discussion, which went on for several days, about the section dealing with the 2000 primary and election:

> **2000 primary/election**
>
> This article has the involvement of many, many editors but I am sorry to say that the 2000 primary section looks badly written to me. I think it needs a complete rewrite.
>
> Why not just cover the facts? That McCain was the major competitor, Mrs. Dole dropped out, Forbes was running. Instead, it covers some disjointed statements about church, conservatives, and Rove.
>
> In WP, I am not for or against Bush, I merely write facts and try to make it sound like an encyclopedia. Is there any opposition for me to re-write it? If so, I'll fuck off. Chergles (talk) 16:53, 11 July 2008 (UTC)
>
> You're right. At least the first paragraph reads more like a newspaper's "funny" section than an election report. If you

rewrite please don't use the "f" word;)—Floridianed (talk) 17:18, 11 July 2008 (UTC)

Support- a rewrite is definitely in order.—SMP0328. (talk) 19:29, 11 July 2008 (UTC)

This would definitely help. Thanks for volunteering. Nishkid64Make articles, not wikidrama) 19:56, 11 July 2008 (UTC)

The 1st draft is done. It's supposed to be factual and neither a smear of Bush nor a glowing report of him. I think the re-write covers the primaries more comprehensively than the previous version of quotes, church, conservatives, Christ, etc. Of course, it can be improved. Overall, I think the length is appropriate to Bush's biography as the primary was a notable event but not worthy of long paragraphs. Chergles (talk) 19:25, 12 July 2008 (UTC)

So far so good. Thanks for your work and don't worry about possible further improvements since you took the first step and a good job. Maybe others might kick-in to help. Thanks—Floridianed (talk) 23:12, 14 July 2008 (UTC)

Chergles is proposing a big intervention—a substantial rewrite of a whole section. Now, he could easily have just gone ahead and edited the page without announcing his intention to anyone. Were there no discussion page, that would be the only option available. But he chose an approach that was less likely to offend those who wrote the earlier versions. Here Chergles tries to clear the proposed rewrite in advance. He recognizes the work of others, but then goes on to point out the article's flaws. Granted, he might have done so more tactfully (and could probably have done without the fairly aggressive sign-off), but overall his intentions are cooperative, and when Floridianed steps in to support the rewrite and soften the tone, tensions are diffused. By the time Nishkid64 chimes in a friendly "thanks for volunteering" and a link to a page, entitled "make articles, not wikidrama," for talking over conflicts, the discussion is civil—the conflict has already been resolved.

These discussion pages have been so popular, and so successful in instilling a sense of community and common purpose, that they soon grew into community portal pages, such as the Village Pump, that allow users to converse about a range of subjects (unlike discussion pages, which are limited to a single subject), including the design of the site itself. These pages not only allow for deeper, more sophisticated conversations, they also include rudimentary social networking features, which, as we've seen, help humanize the communicators and allow them to forge even stronger connections. Registered users can fill out a personal profile—their hobbies, education, likes, dislikes, etc.—establishing themselves as not just an occasional user, but a consistent, human presence. Taking it one step further, a few hundred of the most dedicated and active Wikipedia contributors soon established the annual Wikimania meeting, where they actually meet face-to-face, to get to know one another and discuss the issues most important to the community.

Of course, with the rise of sites like Facebook, MySpace, LinkedIn, and others, features like discussion pages or walls and user profiles have become ubiquitous. But there's a good reason for this: In many cases, the simple ability to communicate and coordinate with others is enough to get people to act together for a common goal, whatever that goal might be. While social networking sites like Facebook mostly started out simply as places for people to write about their favorite books or movies, chat with their friends, or post photos of last night's party (the common goal, in other words, being to form social bonds and seem "cool" to one's classmates), in recent years these goals have gotten a bit more lofty, as social and political movements of all stripes—be it for protests against the policies of the Burmese junta (which garnered more than three hundred thousand supporters on Facebook in a matter of two weeks) or grassroots environmental campaigns—have migrated there. The simple ability to communicate with like-minded others, form bonds over mutual interests, and form coordinated plans is enough to draw motivated people looking to participate in a cause, and get them to cooperate effectively.

What's most fascinating about this trend is that the cooperation fostered by these platforms doesn't just happen online; the relationships and efforts extend offline as well. The best example of this is a site called Meetup.com, on which people living in the same area who are devoted to a common goal or cause can find one another and organize in-person meetings. In addition to being a nice way to meet like-minded people, Meetup has become a particularly effective organizing tool for local movements (my favorite is the small-dog owner's campaign for a special small-dog playground on the Upper East Side of Manhattan—maybe not the epitome of radical political mobilization, but certainly a sign that, to paraphrase Sinatra, if cooperation can make it there, it can make it anywhere).

Communication, of course, is pervasive throughout our lives; there is nothing special about online platforms, except perhaps that they make communication so much easier to coordinate. Any method of lowering the barrier to communication is equally effective, though. For example, one of the first reforms that Toyota made when it created the NUMMI plant in Fremont, California (which I alluded to in the first chapter and will discuss in much more detail in chapter 9), was to change the design of the workstation. Instead of standing alone on the assembly line, workers were placed in constant communication with the other members in teams of five. In addition to these day-to-day interactions, the company also began to arrange for employees to meet periodically over lunch—both to talk about problems or brainstorm quality improvements, and to get to know one another. It may seem frivolous, but countless management studies have shown that these kinds of initiatives lead to measurable results; so much so, in fact, that this kind of team-based production built on continuous communication has become a major part of contemporary management and organizational strategies, not just at the recently beleaguered Toyota, but at some of the other biggest and most successful companies around the world.

One of the most fascinating studies in this area is the work of the renowned management strategists John Hagel and John Seely Brown, who have studied what they call "creation nets." These are loose networks of firms that work together to come up with new products, processes, and ideas. Communication plays a key role in these networks, because in order to work, they depend on the various players in the creation net truthfully revealing what they can realistically do, and at what costs, and what makes sense from each of their perspectives. What's most interesting about this phenomenon is that it works even when members of the net are in competition with one another for some piece of the business or the market. If the predictions of traditional economics were to hold true, this process should fail miserably. The myth of universal selfishness, after all, would predict that members would deceive or withhold information from one another whenever they could do so without detection to give themselves the competitive edge on coming up with the most innovative (that is, profitable) new idea or product. And yet, they do not.

Motorcycles and Mediators

One early example of the kind of "creation net" that Hagel and Brown describe can be found in the motorcycle factories of Chonqing, China. You may not have heard of Chonqing, but it produces more than 40 percent of the motorcycles made in China; no small feat given that China produces more than half the world's motorcycles. As organizational scientist Silvia Pulina describes, the collaborative approach in Chonqing was born in 1990, when a young entrepreneur named Zuo Zongshen began to sell engines he had assembled from spare parts in his repair shop. At the time, the government-owned motorcycle industry was selling parts so cheaply that all the parts for an engine could be individually bought for several hundred yuan less than an assembled one. Zongshen saw an opportunity to make a tidy profit. By the time his practice was discovered by the Chinese government (which tried to put a stop to it by refusing to sell him parts), he already had a backup system in place: He had created a network of three hundred small shops, which would sell him what he needed. These suppliers in turn engaged in a form of cooperative competition. They met face-to-face in teahouses (and later on the Internet) to discuss improvements and cost reductions, and built an informal and completely functional (albeit illegal) business that allowed them to innovate in small ways and make better engines, at lower cost, than the official industry could. The critical piece of cooperative communication is the one that involves conversations of the following nature: "I can lower my cost by *x* percent if you could move that piece two centimeters to the left," to which the response might be "I can move it two centimeters to the left if so-and-so over there can make this pipe more flexible by a factor of *n*," and so on. This kind of interaction can only work in an environment of trust and continuous communication, which then allows the participants to combine these insights to produce the lowest-cost, most efficient motorcycles and share in the benefits of their increased competitiveness. When, in the late 1990s, some of the restrictions on free enterprise were lifted and Zongshen could start selling his own brand, the business flourished. His engines were both low cost and high quality, thanks to the collaborative efforts of the suppliers in his network.

Another real-world example of how communication can facilitate cooperation in inherently competitive situations is the rise, over the last twenty years or so, of alternative dispute resolution in law, in particular mediation. The lawsuit, in particular in the adversarial system that we have in the United States, is designed to resolve conflict by fiat. Irreconcilable parties come to an impartial judge, who wields the power of Leviathan and adjudicates the decision. The parties, at least in the traditional model, are represented by a third party—the lawyer—who is bound by professional ethics to represent his or her client zealously. The lawyers have, in a very real professional sense, a duty to be aggressive and noncooperative and to try to get the very best outcome for their client. (Historically their role was also to be officers of the court and work to preserve the values of the system of law itself—for example, not to lie to the court, not hide evidence, etc. This seems to have been honored more in the breach than in the performance, but that's another story.)

Litigation turns out to be a very expensive and not particularly satisfying approach to conflict resolution. In its stead, a movement has emerged within the legal profession that is pushing toward mediation. The hallmark of mediation is simply for an impartial third-party mediator to get the parties talking to each other, face-to-face, to try to understand their respective needs and constraints, and to work out a resolution they can all live with. Reading through some of the handbooks on mediation immediately gives one a sense that this process is pretty much as nonadversarial as you can get (at least where lawyers are concerned). *The Mediator's Handbook,* by Friends Conflict Resolution Programs, begins thus: "Two mediators . . . open with a welcome and an explanation of what will happen. Each person takes a turn speaking while everyone else listens. For the most part, this is open-ended: the person can talk briefly or at length about anything relevant to the situation." Later: "the mediators do not try to determine the truth, or who is at fault. Rather, they listen for what matters to people and for possible areas of agreement."

In other words, *communication* between the parties, relatively unstructured, exploring what each cares about and how they view the situation, is the foundation of the process. The mediation handbook prepared a few years ago for the Harvard Negotiation Program specifically describes communication as one of the pillars of mediation. Another handbook lists among the skills of the trained mediator, "Understand the importance of demonstrating empathy, building rapport, establishing trust, setting a cooperative tone, demonstrating neutrality and impartiality, demonstrating sympathetic listening and questioning."

The basic point is clear. Mediation is a model of conflict resolution that replaces the iron fist of the Leviathan with mutual recognition, understanding, and discussion. Many of the foundations of a successful cooperative system that we discuss in these chapters—empathy, fairness, and trust—repeatedly show up as core components of this process. But communication is the most fundamental one. The most skilled and successful mediators, though, are those who not only foster the most effective communication, but also are able to frame the conflict in such a way that both parties come to see a fair resolution as more achievable than they previously thought.

CouchSurfers and Zipcar Drivers: Cooperation and Framing

In looking at mediation, and at the other examples of successful cooperative systems, we begin to see why it is that communication works as well as it does. First and foremost, practically every one of the elements of a successful cooperative system I discuss in this book—empathy and solidarity, moral norms, fairness, trust, and leadership—depends on communication. The difference between an anonymous, mute interaction and one in which we are able to communicate is the difference between the artificial setting of the lab and the real, rich world of human interaction.

In addition, communication helps us define the situation we are in, or frame our interaction. As we saw in chapter 3, the human psyche and brain do not respond to a set of facts that come without some form of interpretation. Our experience of any situation is determined not only by the objective facts of that situation, but also by all the subtle cues and associations that go along with it and give it social meaning. Our interpretation of any given situation is shaped by the way the facts of the situation are communicated.

Seeing how we could reframe situations to get people to cooperate in the real world is in some senses simple, and in others very challenging. Saying the right words may be easy; actually communicating them in an authentic and credible way is harder. But several examples show that it can be done. Take for one an online phenomenon called couchsurfing.org. CouchSurfing is a site used by half a million people around the globe that connects travelers who need a place to stay (a couch) in a foreign city with people willing to host them. This isn't a room rental service, by which people rent out their rooms (or couches) for a fixed or mutually agreed upon fee; couchsurfing.org actually *forbids* payment (though it does encourage surfers to bring gifts, or help out around the house, or offer some other kind of nonmonetary compensation). Now, you might be thinking this arrangement makes sense for the guest—who wouldn't want a free place to stay in a foreign city? But what incentive does the host have to offer up his home, for free, to a complete stranger (who could easily be an ax murderer, thief, or weirdo)? Conventional wisdom would assume that the site would have countless would-be travelers but a severe shortage of hosts. But this turns out not to be the case. Why? Because, just like the Kiva site that we discussed in the last chapter, CouchSurfing, which has all the trappings of a social network, has framed itself not as a business but as a community of people with shared interests and values—such as travel, learning about other cultures, and so on. Not only does it frame itself as a community by stressing these shared set of values, it also communicates a strong set of community norms, such as reciprocity (most hosts will at some point be guests, and vice versa, so it's technically indirect reciprocity), communication, and trust. The site relies heavily on a small number of engaged leaders (who both founded the site and continue to use it both as CouchSurfers and as hosts) to instill these norms (such as the ban on payment) by being in continuous contact with members in the form of postings, emails, and the site's social networking features. The founders of the site keep the system (and the cooperation it depends on) running smoothly, in other words, not by merely saying "this is a community" but by taking authentic measures to sustain the *feel and experience* of an actual community. And these results can been seen in the stories that people post about their CouchSurfing experiences; in reading the message boards, one does have the feeling that one is observing a true community at work.

Now, CouchSurfing isn't a for-profit enterprise. But as it turns out, even businesses and companies can benefit from framing something as a community rather than a strictly for-profit enterprise. One successful example of this is Zipcar, a "car-sharing" company that originated in Boston. Zipcar's business model is simple. For a small membership fee, people can rent a car by the hour, for relatively low rates. The draw of Zipcar, though, is less price than convenience; the company keeps fleets of cars in central locations where renters can easily pick them up and drop them off. But this is not the only reason for its appeal. Like CouchSurfing, the company forges a sense of community among its users. How? For one, they create a sense of solidarity and shared purpose by targeting a certain type of customer: the environmentally conscious. Their advertising and website tout their corporate mission—to help reduce pollution by making it easier for people to avoid owning a car. And most of their cars support that mission; they are small and fuel-efficient (many are hybrids). Like CouchSurfing, Zipcar frames the transaction as being at least partly social (though of course money does change hands) by calling it car "sharing" rather than rental. A strong element of trust further reinforces the community feel; there are no attendants to check the condition of a car once it is returned; members are simply encouraged to return the car clean, in good shape, with a full tank of fuel, for the next user—which they do, with amazing consistency. In other words, the company presents itself less like a business and more like a sort of environmentally conscious club.

Is this a manipulation? Not really. To listen to Robin Chase, one of the company's cofounders, there's no question that the company really was founded on a belief of doing well by doing good: of improving urban transportation while reducing carbon emissions. But the pureness of intent isn't the point; the point is that painting itself as a community not only wins Zipcar customers, it also motivates them to follow the rules and treat the cars, and other members, well. In this, it is like a real-world version of the Wall Street/Community Game experiment. The big question that remains, of course, is how far this can go, and at what point customers begin to look (as I think many now do) at the overuse of the word *community* by businesses as sheer manipulation. This, needless to say, does not undermine the importance of framing. It just means that the frame has to be communicated in a credible way, and that one company's authentic claims can be devalued by another's manipulation.

The basic point of all this is simple: Nothing is more foundational to cooperation than communication. Talk is not cheap; through it we can come to define our preferences, goals, and desires in a situation; begin to build mutual empathy; negotiate what norms are appropriate and what course of action is fair; and begin to build trust and understand one another. Communication is the one thing that has the most unambiguous, most dramatic effect on cooperation, both in experiments and in the world. It is the one thing that exists wherever cooperation is successfully practiced—be it motorcycle makers in the teahouses of Chonqing, mediators in American courtrooms, or globe-trotting travelers looking to share a couch in one remote corner of the world or another. Cooperative systems across the globe have one thing in common: They all depend on communication.

• CHAPTER 6 •

Equal Halves: Fairness in Cooperation

There's an old Yiddish joke about Herschel of Ostropol, who is walking with a friend when they come across a cookie left by someone on a bench. Herschel picks up the cookie, breaks it into two pieces of unequal size, and hands the smaller piece to his friend. The friend looks at him and says, "Herschel, what are you doing?" Herschel answers, "What's wrong? What would you have done in my place?" The friend answers: "Well, had I broken the cookie like that and there were two unequal pieces, I would have taken the smaller one for myself, and offered you the larger piece." To this Herschel answers: "Well, what are you complaining about? That's exactly what I did!"

We care about fairness. Even young kids usually have

fairly honed senses of justice ("I should get as much as she's getting," etc.). Parents encourage this by giving siblings equal gifts, clothes, allowances, and so on. Later in life, we usually develop more complex notions of what is fair. Other facts enter into the equation, such as relative need, luck, and talent. We come to accept that some people are better off than others, and that's just how it is. And yet we still care about fairness in one way or another. What, then, might we be caring about when we feel that we care about fairness?

In looking through the experimental economics and social psychology literature, it seems that when we care about "fairness" we really care about three distinct things: fairness of outcomes, fairness of intentions, and fairness of processes. With regard to outcomes, we care about how much each of us gets out of an interaction relative to others, given the generally understood norms. For intentions, we particularly care when the outcomes are not "fair" given generally understood conventions for the situation, whether the unfair outcome was intentionally brought about or not. And as for processes, we care whether the way in which the outcome was achieved was fair or not, whatever the outcome and the intentions of the people involved.

Let's start by looking at the evidence for fairness in outcomes. First, we need to understand that caring about fairness, as opposed to simply about how the outcome affects me, is a deviation from pure self-interest. If I would rather get, say, $10 while you are also getting $10, than get $12 while you get $100, then you can say that I care more about fairness of outcomes—how equal our shares are, in this case—than I care about my total payoff (which is higher in the unfair distribution than in the fair one). In a series of ultimatum game experiments, Swiss economist Ernst Fehr and several of his collaborators—Klaus Schmidt, Urs Fischbacher, and Armin Falk—demonstrated that indeed, under controlled experimental conditions, we do care about the fairness of the outcomes completely independently of what we will come away with, sometimes to the point where we would rather walk away with nothing than agree to an unfair share of a deal.

The workhorse experiment in this line of literature is the ultimatum game. In this game, the experimenter gives one participant, the proposer, some amount of money, say $100. The proposer can then give whatever portion of this he wishes to the other participant, the responder. If the responder accepts the amount, each goes home with what they agreed upon. But if the responder rejects the offer, then both go home with nothing.

Imagine you were playing this game and the proposer offered you some pittance, say, $1 of his $100. How would you respond? On one hand, a dollar is better than nothing. On the other hand, you think, "What the hell? This guy is being selfish! We're both here as experimental subjects, he got lucky, why can't he share more fairly! Screw him!" Sure, a dollar is technically better than nothing. Standard game theory would predict that respondents will always accept whatever is offered to them, because going home with something is better than going home with nothing. But knowing that the proposer had $100, most people would be tempted to punish him for his stinginess and reject the offer, leaving him with nothing. In experimental settings, this is exactly what happens, and in fact people reject offers much higher than 1 percent. Countless studies conducted across several countries show that half the time, responders reject offers that are less than 20 percent of the total amount. Proposers seem to expect this, which is why, at least in industrialized societies, a majority of the offers are over 30 percent. In fact, by far the most common strategy of proposers is to offer a fifty-fifty split, to ensure that the responder will accept. From the perspective of rational actor theory, rejecting any offer greater than zero is not in our self-interest—it basically amounts to, as the old saying goes, cutting off our nose to spite our face. But this logic ignores the simple fact that we really do care about being treated unfairly. So much so, in fact, that we are willing to punish others for acting unfairly, even when it costs us something in return.

Just as subjects were quick to punish those who acted unfairly in the ultimatum game, people are equally quick to punish others for unfair behavior in real-world situations as well. This willingness to punish, even at a cost to ourselves, can be very beneficial to a society in that it helps enforce cooperation (although the effects are complex, and not always positive, as we'll see in chapter 8). And, as it turns out, it actually has a biological basis. In one study, Fehr's group teamed up with neuroscientist Dominique de Quervain to try to "see" the effects of punishment at work in the brain. When they scanned the brains of subjects who were given the opportunity to punish other players in Prisoner's Dilemma games, they found that those who spent more money on punishing defectors had more activity in an area of the brain associated with rewards, or pleasure. In other words, certain people got an internal, neurological high from punishing those who had wronged them.

If we agree that people care very much about fairness, the next question to ask is what constitutes being "fair." This is not an easy question. There is no single clear definition of justice, or fairness. The simplest way of defining fairness is equality—everyone getting the same percentage of the pie. This is probably why Elinor Ostrom found that property-sharing, or "commons" systems (which we'll talk about in greater detail in the next chapter), work better when everyone has a very similar endowment and set of needs: similar or equal land, access to water, or roughly the same number of cows grazing on a common pasture, et cetera. But achieving such a neat, equal distribution of resources isn't easy in most situations. What if the size of two people's land is equal but one is in a better location and thus is higher in value? Or what if people have access to the same amount of water but one has thirstier crops to irrigate? Such examples show us (and in fact these are still relatively simple compared to most real-world situations) that we usually will require, and be willing to live with, more complex definitions of fairness.

Gold Miners, Shipwreck Sailors, and Politicians: Fairness of Outcomes and Intentions

Fairness can mean quite different things for different people in different settings. One example of this comes from the work of legal historian Andrea McDowell. McDowell studied the mining codes developed during the 1848–49 California gold rush. Because camps were cropping up like mushrooms after a rain, miners were transient, and the territory had not yet been formed into a state, it was impossible for authorities to effectively enforce a single, formal property law over mining rights. So instead the miners in each camp set up codes themselves to ensure that the distribution of the land would be more or less fair. They all agreed it was unfair for any one miner to claim more land than he could work, but beyond that, rules differed widely across the camps. Some camps allowed miners to buy and sell lots; others didn't. Some allowed miners to own several lots as long as they hired people to work them; others prohibited any miner from holding more land than he could physically work by himself. Whatever the rules, the point is that even within these small, insular communities of miners, there was a range of notions about what fit within the broad definition of "fair."

When we look beyond industrialized societies, we find that the range of definitions of fairness is indeed enormous. In a groundbreaking study, anthropologists Rob Boyd and Joe Henrich and economists Sam Bowles, Colin Camerer, Ernst Fehr, and Herbert Gintis conducted ultimatum games with subjects from fifteen small-scale societies all over the world. They found enormous differences in conceptions of fairness. Some people, such as the Machiguenga in the Peruvian Amazon, perceived highly uneven distributions as fair; they tended both to offer low amounts and to accept low offers from other players. This behavior makes sense within the cultural context of the Machiguenga: they live in small, self-sufficient kin groups and almost never transact with others, so it's understandable that they have no expectations of fair exchange with outsiders and treat any gift from a stranger as better than nothing. People from cultures like the Anguganak and Bogasip of New Guinea, the researchers found, placed high demands on "generous" giving. They tended to make a wide range of offers, but also had much higher rates of rejection; some responders even rejected supergenerous offers of up to 70 percent. This too makes sense given their culture of gift giving. David Tracer, who conducted the New Guinea study, suggests that in these societies gifts always came attached to high and uncertain social obligations; rejecting a hypergenerous gift may reflect an unwillingness to accept the obligations that were thought to come with the gift. In both societies, no standard of equal division existed—in the former, because cooperation with strangers is such a rarity that robust fairness norms did not develop; in the latter because gifts play such a large role in social obligations.

• • •

Given these differences across cultures, the patterns of behavior in ultimatum games in developed societies, which are centered around equal division, become less surprising. We are used to interacting with strangers in mutually beneficial exchanges all the time. We are used to trying to follow simple rules for the division of benefits. We are used to thinking that we should treat one another as more or less equal. We follow simple rules that assume the social roles we inhabit—customers, vendors, employees, and so forth—to be equal. And so when we find ourselves in an experiment meant to elicit our beliefs about what constitutes fairness, we revert to these notions: equal division.

Our expectations about fairness also depend on our expectations in a given situation or context. You might think it's perfectly fair for the top 1 percent of all earners in the country to pay 50 percent of their income in taxes and have it redistributed to the poorer members of society. But imagine that you were in Las Vegas, at a casino that taxed all winnings by 50 percent and then redistributed the money to that evening's losers. Would you still think it fair? Or you might believe it's perfectly fair for someone who is a good negotiator to get a better deal on a used car than someone who isn't. But what if someone other than the highest bidder at a public auction walked away with the antique chest? The difference here is not the equality of the outcome itself (all outcomes are uneven); it is in our expectations in that context. In terms of government programs and revenues, we expect that relative need will be factored in. In consumer contexts, we expect to be taken for every penny if we're not careful (hence the popularity of the phrase "buyer beware"), although many companies strive to overcome that oppositional relationship. In each situation we have very different expectations, and when outcomes don't meet those expectations, we perceive them as unfair.

This is really just another way of saying that we are far more willing to accept or tolerate unequal distributions of wealth or resources when they are presented, or framed, in certain ways. In ultimatum games, when responders are told that the other player either "won" their endowment in a lottery (in other words, through luck) or "earned" it by answering quiz questions correctly (that is, through skill), responders tended to accept lower offers than when they were told nothing about the origin of the money. Again, this is perfectly consistent with how we tend to justify inequality in our society. We live in a culture that tells us that we should not begrudge the wealth of people who are diligent, creative, and productive; we often describe it as a fair reward for their superior skills and efforts (though we might feel envy). Interestingly, though, we consider distributions to be "fair" when the gains are won by being lucky.

This is particularly true in the critical lottery of the gene pool—of being born to the "right" parents. We don't see luck as being unfair; we see it as just a part of life. At least in American politics, part of the resistance to redistribution policies comes from the widespread cultural commitment to the idea (one that is not, by the way, empirically grounded) that most inequality is deserved, that is, fair. We assume it is mostly based on productivity, hard work, et cetera, and that even the accident of birth counts as a fair basis for getting an unequally large opportunity, because it is seen not as the unfair advantage of the child, but as an expression of the deserved well-being of the parents who successfully care for their children.

So "fair" doesn't have a universal, stable definition; it varies according to both cultural norms and the situation involved. It is also subject to individual interpretation. Imagine the following scenario. You are a ship captain and you have just picked up a boat full of shipwrecked survivors. You have a limited supply of water. You know how long it will take you to sail to safety and have calculated how much water you can afford to give to each of your crew and to the survivors you picked up so it will last long enough without anyone dying of dehydration (although everyone will be quite thirsty). However, after a few days, and with several more days to go, a couple of the shipwrecked sailors seem to be in very bad shape: They are weak, beginning to develop terrible sores, and are moaning incessantly. What do you do? Do you give them slightly larger allotments of water, or do you stick to your original plan? A captain operating under the narrow definition of fairness might enforce the allotted division ruthlessly, ignoring signs of extreme distress in individuals. But a captain who feels it's unfair for some of his crew to suffer more than others (even if they are getting the same amount of water), on the other hand, will not be able to bear the misery of the visibly suffering sailors and will give them extra water, even at the expense of the other crew members and himself. Which is the right thing to do? One could argue that both are technically fair; the first is treating each member of the group equally, and the second is taking into account their relative need. The story was originally introduced into the psychology literature by Daniel Batson as a way of probing when empathy might actually be counterproductive—the captain who is too empathetic might endanger the lives of the whole crew if the water runs out before they reach safety. But it also underscores very powerfully that different conceptions of fairness can lead to radically different distributions, all of which could be justified in context, and each of which could have very different implications for everyone involved.

The basic psychological and behavioral fact that we care about relative need creates a real challenge for American politics. Western European social democracies respond to it through their progressive tax and transfer system. In the United States, the emphasis shifts to equality of opportunity: That is to say, American political culture is built on the notion that everyone should have an equal chance (or so the story goes in theory, if not always in practice). Similarly, the very strong American emphasis on the spirit of entrepreneurship, individual achievement, and pursuit of wealth reinforces a theory of fairness based on effort, talent, and contribution rather than strictly equal outcomes. This is in part why proponents of health-care and welfare reform try to reframe the debate, using categories such as children, the elderly, and those born into disadvantaged circumstances as groups needing of special protection, so as to justify the redistribution of benefits without upsetting the core narrative about fairness in the United States: a fairness of process, or opportunity, rather than a fairness of outcomes.

We have a strong concern not only with fairness of outcomes—who gets what and how much—but also with fairness of *intentions*—whether or not someone is deliberately being unfair, or selfish. Just as we are more accepting of gains won through luck or skill, we are more tolerant of outcomes that are unfair for reasons out of anyone's control than we are of outcomes that are unfair because someone is taking advantage of the situation. Scientists have attempted to tease out how we respond to intention versus how we respond to outcomes through a clever manipulation of ultimatum games and similar experiments. In such experiments the computer determines how much the first-moving player will offer his partner. Indeed, when responders know that the other player has no control over how much he or she offers, the responders tend to accept lower offers in ultimatum games.

Not only do we tend to forgive unfair outcomes that are out of another person's hands, but we also tend to be more understanding, even sympathetic, when another person puts us at a disadvantage, as long as the only way they could have avoided doing so was by imposing a large cost on themselves. One experiment showing an example of this was conducted by Armin Falk, Ernst Fehr, and Urs Fischbacher in which they gave one group of proposers the option of choosing between a 50:50 split and an 80:20 split; the other group could only choose between an 80:20 and a 20:80 split. Although the outcome when 80:20 was chosen was of course equally unfair, fewer responders in the second group, where the only alternative the proposer had was to take 20:80 themselves, rejected the 80:20 split. Perhaps no less interesting is the fact that some subjects were so resistant to being treated unequally that they rejected offers of 80:20 even in the game where someone *had* to lose out. These people seemed to think, like Herschel's friend, that when luck hands two people a highly unequal distribution, the person who has the power to choose between the options must take the short end of the stick for themselves.

In another fascinating study, Tania Singer and her collaborators ran sequential Prisoner's Dilemma games and then presented the subjects with images of the faces of the people with whom they had played. During the game part of the experiment, the first player made a move, which could be cooperative or noncooperative. The worst payoff for each player, as with all Prisoner's Dilemma games, occurred if one player cooperated and the other defected. In the sequential Prisoner's Dilemma games, if the first player defected, then obviously the second would defect, too. But if the first cooperated, then the second player had a real choice: cooperate, which would make both players better off, or defect, which would be even better for the second player but worse for the first player. Sometimes the second player had an opportunity to choose whether to cooperate; in others the player had no choice and had to follow the directions of the person running the experiment. As a result, some of the second players responded cooperatively, others not, and some acted intentionally, while others did not.

Next, Singer showed the subjects pictures of the people they had played. First, she asked them whether they remembered the face. Second, she asked how "likable" the person was. And third, Singer scanned the brains of the subjects. What the researchers found was that the areas of the brain activation involved in both evaluation and reward were activated differently, depending not only on whether the second player had cooperated, or behaved fairly, but also on whether the second player had done so of his or her own free will. The subjects also remembered the faces of cooperators who did so intentionally more clearly than those who either didn't cooperate or cooperated but had no control over their choice. Their assessments of likability followed the same pattern. Essentially the experiment showed that we remember and respond favorably to people who cooperate, we remember people who intentionally defect (and have a negative attitude about them), and we don't remember or have much of an opinion about people who hurt us (in the game) unintentionally. This makes sense. As a mechanism to navigate through life, it's beneficial to us to be more easily able to remember people who either intentionally cooperated with us or intentionally harmed us, in terms of knowing whom we can trust in the future. There's no reason to believe that a person who harmed us *un*intentionally will do so again in the future; it's fine if they are fuzzier in our minds. Fair intentions matter, then, independently of fair outcomes.

Lotteries, the Draft, and Why We Wait in Lines

Of course, while intentions and outcomes are easy enough to tease out in experimental settings, in real life they are generally more difficult to unlink; after all, a situation in which someone is deliberately trying to maximize his own gains at another's expense is unlikely to result in an outcome that anyone would consider fair. This is why, for those interested in designing a fair system of any kind, it is helpful to look at the third dimension of fairness that we care very much about—the fairness of processes. The fact that participants are willing to accept less equal splits when the proposer has "won" or "earned" the endowment suggests that if we deem the process that produced an unequal outcome—such as a lottery, or an executive bonus award process—to be fair, we'll accept the outcome as a fair one even if the outcome is unequal.

There is evidence to suggest that we care even more about the process than the actual outcome of a situation. How many times have you heard someone accept a less than favorable outcome with the words "what's fair is fair." One type of process that we widely perceive to be fair, regardless of outcome, is one that is random, again, like winning the lottery. Take how the public perception of the Selective Service draft process shifted during the Vietnam War. At the start of the war, there were all kinds of deferments—for education, fatherhood, certain professions—that clearly favored some (wealthier) groups over others. By 1969, however, as resistance to the war grew louder and stronger, most of these deferments were abandoned and a random lottery system was implemented. This didn't make the war any more popular, but it was clear that as the pressure mounted, the government had to do something to blunt the rage about the terrible unfairness of the fact that some young soldiers were sent to fight and die, while others weren't. To do so, it resorted to a process of blind luck rather than a more rational process based on reason (one could argue that it's more reasonable that someone who is, say, training to be a doctor or a physicist be allowed to stay behind because he or she is ultimately performing a greater service in their profession than they would as a soldier).

This perception of the lottery process as being inherently fair dates as far back to biblical times. In the Old

Testament, when the Israelites need someone to blame for angering God and costing them failure in conquering a town, they draw lots to decide. Similarly, in the book of Jonah, Jonah is chosen as the scapegoat for a storm because he drew the unlucky lot.

One would think this kind of scapegoating would be perceived as wildly unfair in modern culture. But in reality it's not so different from modern-day practices like the draft, or the lottery for Super Bowl tickets among season ticket holders. In these cases we all agree that the outcome isn't especially fair (especially for the season ticket holders who don't get to go to the Super Bowl), but because the process is random, the overall system somehow is seen as just.

But not all processes can realistically be random—if they were, our world would devolve into chaos. Think about what happens when you arrive at the ticket window to the subway, or a theater. Where do you stand? Stupid question! At the end of the line. It would hardly be practical for everyone to huddle in a big mass and wait for the guy behind the window to pick a name out of a hat. Even so, "first come, first served" isn't quite as simple as it sounds; in order to be effective, there has to be some easy way of signaling who came when. Hence the well-understood signal of the line—we all know that when we arrive somewhere, we go to the end of it and wait our turn. The line also makes the first-come, first-served rule incredibly easy to self-regulate; we've all seen what a crowd of people will do to someone who brazenly tries to cut a line. It might not be based on merit, nor might it be completely random, but in our culture, "first come, first served" is widely perceived as a fair process.

As things get more complicated, the fairness of processes becomes no less important, though defining what makes a process fair can become harder. Think of the image of Justice, blindfolded with her scales. She needs a lot of help, in the form of many layers of policing and judicial procedure, to maintain a sense that the process over which she presides—the law—is fair. The most extensive research on the fairness of procedural justice in the United States has been done by psychologist Tom Tyler, who studied attitudes toward the law in poor communities. Surprisingly, he found that the response of people in these communities to law and law enforcement had more to do with the perceived fairness of the processes than the outcomes. All things considered, people who had been handed the short end of the social and economic stick still saw law enforcement as legitimate, and law-abiding behavior as necessary, when they perceived that law enforcement as a process was fair.

Truckers and Cement Mixers: Defining Fair Pay

So what does fairness look like in a work setting? Let's consider an obvious example—compensation. In some people's definition of a fair compensation system, wages are based completely on merit—either effort or skill. For others, equal pay for people with the same seniority is seen as the appropriate measure. For still others, fair wages are those determined solely by the market—supply and demand dictate the value of someone's time and work. Indeed, traditional economic modeling would predict that markets determine wages, period; fairness has no place in the labor market. In theory, if one firm pays much more than others do for similar work, it will attract more workers than it can accommodate. That should drive down the wages it pays, as long as it pays more than its comparable competitors. If for some reason it doesn't, then its labor costs will be too high relative to competitors, and it will go out of business. In theory, therefore, a firm that wants to pay wages that, for example, give workers a fairer share of the firm's profits simply will be forced by the costs involved to pay less. The market, not fairness, will determine wages. Reality (as in many other cases we have seen throughout this book) refuses to comply with this model, however. In actual studies of wages in markets there are huge disparities among firms within a single industry, often with as much as a 50 or 60 percent difference between what different firms pay for comparable jobs and workers. As a result, we have a good bit of variation within industries, which allows us to look at firms that offer wages that are more or less fair along various dimensions.

The easiest comparison is between firms that use a definition of *fairness* that means "merit," or pay-for-performance, and those that use *fair* to mean "equal," as in having fairly flat wage structures. Management studies have looked at a range of firms: some that have flatter wage structures and others that are more focused on pay-for-performance. Each approach has successful examples; each has its failures. Having a very flat pay scale has the obvious benefit of equal outcome—everyone (at least everyone in a certain position) gets paid the same amount. But it also means that workers who perform poorly and slack off still get paid as much as those who are working hard and making real contributions. Remember, fairness is not only about equal outcomes, but also about outcomes that fit accepted norms of fairness.

Take, for example, a set of studies conducted in the trucking and cement industries by Jason Shaw and his collaborators. The studies attempted to map the relationship between wage disparity and performance. They defined performance using both objective measures, such as speed of delivery or number of accidents in the case of trucking, and hours of labor per ton of cement in the case of cement, and subjective measures, such as how the employees were perceived by their supervisors. In both industries, one thing was clear. Employees in companies where wage disparities were high but were closely tied to performance got high marks; they perceived effort and performance to be a fair criterion for setting pay and so they responded by working harder. In companies with high wage disparities that weren't tied to worker performance—that were instead the result of chance, nepotism, or other factors that were not seen as fair—employees performed poorly. And why shouldn't they, if they knew that hard work wasn't going to be rewarded, and that the goodies to be had went to people who didn't deserve them?

But what about employees in companies that offered no incentives for performance at all, and paid equal or flat wages? If we are all driven by self-interest, the assumption would be that people universally perform better when given material incentives, right? But in fact, in these studies, people performed equally well under a flat pay scale as under incentivized pay-for-performance, if that flat scale was the formal policy of the company. And they performed poorly if the company described itself as rewarding effort with pay, but then paid a flat scale in reality. In both cases the expectation, or the normative framing, was critical; people performed better when the actual pay fit the widely shared norm for the company (equal pay or incentive payments). What *didn't* work were payment schemes, such as nepotism or ones involving other unfair advantages, that either did not fit the stated policy (or norm) or didn't fit any widely agreed upon standard of fairness.

Looking at these experiments and field observations, we are left not with a single answer as to what kinds of compensation systems—be they in business, in government, or even in our own households—are most fair. But what they do tell us is that the long-held notion of what it takes to "get the incentives right"—in other words, to impose systems that try to monitor performance and pay exactly for what is done—misses the basic insight that fairness matters. There is no one right answer for what is most fair. But the one thing that is clear is that compensation, as well as other, less material rewards, needs to be distributed in ways that fit our widely shared cultural beliefs about what is fair and appropriate in that specific situation or context. In other words, flat payment schemes are preferable in settings where equal division is expected, and merit-based systems are appropriate in situations where seniority, effort, or performance are emphasized. And when designing such systems, we need to remember that fairness has to do not only with outcomes, but also intentions. Participants in a system or the people who run it (in the above example, those dispensing the compensation) must be seen as adhering to a fair process. We can accept many things as being fair that result in unequal outcomes, from lotteries to some levels of executive bonuses. But not everything goes. There are very few situations in which whim would be seen as a fair process for deciding anything. And as we are learning with the still unfolding public response to the financial collapse of 2008 and its aftermath, there are certain levels of high payments to executives where the sheer size of the payment itself makes the outcome, the process, and the intentions seem unfair and unacceptable.

The critical lesson from work in experimental economics and social psychology, as well as in business and social studies, is that our desire for fairness, as it is understood in a cultural context, is a critical component of human motivation and behavior. It is independent of self-interest, empathy, or solidarity. We have a basic need or desire to be treated fairly, and to be participants in systems that treat us fairly. If we want to build a system that motivates people to work well, or cooperate effectively, it's not enough to offer them simple rewards and incentives. We also need to think about how fair the system is. Fairness is a practical, integral part of making systems work well, and of making people function well and cooperatively within them.

• CHAPTER 7 •

WHAT'S RIGHT IS RIGHT—OR AT LEAST NORMAL: MORALS AND NORMS IN COOPERATION

It had been a long drive up the mountains. My kids, my wife, and I were vacationing in the Pyrenees, on the border between Spain and France. Happy to find any place still open to eat, we were pleasantly surprised to stumble upon a fabulous restaurant where we had the most delicious dinner, served by possibly the friendliest and most professional waiter we have ever had the pleasure of encountering. Few American tourists passed this way, so he was as happy to spend time chatting with us as we were with him. At the end of the meal, we were so pleased with the service, and in fact the overall dining experience, that we left a substantial tip. But to our surprise, when he picked up the tip, his warm expression turned into one of hurt and pain. I immediately realized my mistake—that here, for this person and in his culture, tips were for hirelings, not self-respecting professionals. We had clearly offended his professional sensibilities. So I waved toward the kitchen and the busboy and said, "It's for them, so they can have drinks after work," whereupon the waiter turned around, visibly relieved, and called out to the people in the back, "Drinks!" We parted with warm smiles and goodbyes.

Was what we did wrong? We thought we were being generous. And indeed for us, as Americans, to add a substantial tip to the bill as thanks for a wonderful meal and experience—to reciprocate generosity with generosity—is the norm. Yet even in countries in Western Europe, where the culture is not so dramatically different from our own, the range of norms is wide. In this case, no real damage was done. But these kinds of discrepancies in what is considered normal behavior become problematic if we want to build systems in which people can collaborate across societies and cultures.

In everyday life, norms can be arbitrary. Often it is not the standard itself, but the coordination of standards that is important. After all, a traffic law requiring people to drive on the left side of the road, for example, is not intrinsically better than one requiring we drive on the right. It is, however, clearly better that everyone drive on the same side. Similarly, it doesn't really matter in what direction a line forms; it only matters that everyone joining it know which end is the back and which is the front, that you join a line at the back. In other words, there are many cases where the nature of the norm is irrelevant; what counts is that it is known and followed by all parties, and ideally that it is simple to understand and simple to monitor so that errors are relatively rare and defections easy to observe.

Standards can go well beyond just coordinating people's behavior; they also serve to encourage people to behave well toward one another. Some of these standards or norms are experienced as moral commitments. They are so deeply held that we feel we must act on them even when no one is watching, or even when they force us to put aside self-interest, convenience, or the opinion of others. In our day-to-day lives the norms and mores we follow are like laws, practical rules of thumb, unchallenged habits, and socially enforced practices (that is, acting according to what society does and thinks is right). Just as it is important to us to be seen as fair and benevolent, there is much evidence to suggest it is extremely important to us to be seen as socially appropriate—perhaps even more important than it is to obey authority or the law. In 1991, legal scholar Robert Ellickson published a brilliant book, *Order Without Law,* that looked at how a community of ranchers in Shasta County, California, handled the problem of trespassing cattle. He found that the ranchers blatantly ignored the laws and parameters established by California law, and instead followed the set of informal, mutually agreed upon rules that dictated everything from how long one had to go pick up one's stray animal from another's land to what sort of compensation, if any, was due for which kinds of damage caused by a trespassing cow. In some situations, ranchers were more respectful of their neighbors than the law required; in other situations, they were less so. The point is that order prevailed; it was simply that instead of it being the rule of law, it was the standards of the community that the cattlemen followed. And even though none of these norms was legally binding, they were rarely, if ever, broken. His book launched a whole line of research, as scholars looked at the norms of everyone from grain traders to standup comedians.

Music Downloads: The Power of Combining Fairness with Conformity

One fascinating aspect of norm-driven behavior is that people seem to care a lot about conforming with others about what is seen as "normal" behavior. When we design systems of cooperation, we can use that tendency as a way of encouraging people to choose prosocial behavior. The tax experiments in Minnesota and Australia that I described earlier are very real and powerful examples of how the power of the inclination to be "normal," like others, can encourage us to conform to socially valuable behavior.

A more recent example of this power comes out of a study that I did with Leah Belsky, Byron Kahr, and Max Berkelhammer. We looked at several online music sites that largely relied on voluntary payments. The question of how to get fans to pay for music downloads online has vexed the music industry since the mid-1990s. It has been the single most powerful force behind the increasingly extensive and aggressive copyright policy of the United States. The recording industry has been pushing the U.S. government to adopt ever-more Leviathan-like strategies, from the encryption of music tracks to the prosecution of teenagers. And yet fans continue to share music illegally, and record labels continue to sue them with little to show for it in terms of stabilizing their business model. However, some artists and online labels have been experimenting with less aggressive approaches. They have deployed instead a range of strategies that fit the basic design approach I have been describing in this book to get fans to do the right thing—to voluntarily pay artists for the privilege of enjoying their music, so the artists can make a living and continue to dedicate themselves to making the music we love. Americans spend more on tips than they do on recorded music, which begs the question of how we might find a way to get fans to recognize that their responsibility to the artists who enrich their lives is at least equal to their responsibility toward their café server or taxi driver.

There have been quite spectacular experiments by established artists such as Radiohead and Nine Inch Nails, who successfully released albums under a voluntary payment system. For our study, we chose to look at artists who were not as well-known, and so whose experiences were more representative of the normal artists' prospects. One of the sites provided a particularly powerful example of how simple conformism plays a large role in getting people to make voluntary payments. Magnatune is an online label that sells music of many artists, in each case releasing the music in a perfect digital format that users can copy flawlessly. They also release the music under a Creative Commons license that makes it perfectly legal for the fans to make as many copies as they wish. In other words, when a fan buys music from Magnatune, the fan can *legally* and easily make millions of copies and blanket them across peer-to-peer networks. And yet they don't.

After scouring the records from more than 75,000 transactions on the site, we found that although users were invited to pay between $5 and $18, at their discretion (in increments of 50 cents), 48 percent of users paid $8 per album, well within a range of what the industry would be thrilled to establish as standard practice for consumers. Even more incredible, only 15 percent paid the minimum (without copying from a friend) of $5 and another 15 percent paid as much as $10. What was most fascinating, though, was the distribution of payments. The drop-down menu indicated how much most people paid. Next to $8 the menu said "Typical." Next to $10: "Better than average." Next to $12, "Generous," and so forth, escalating praise in $2 increments until reaching the maximum of $18 and the words "We love you." The power of the prompt is clear: 48.05 percent of users paid $8; only 0.34 percent paid $8.50;

and 2.93 percent paid $7.50. People are looking for a signal that will tell them what is normal—not just what is normal on average, but also what is normal for a person who wants to be seen as "better than average," or "generous," and so forth. My point is, norms matter. People seen to flock to behavior they consider "normal." So if we want to encourage good social habits, we need to do more than institute norms; we also need to set clear signals of what counts as normal and appropriate behavior.

Spanish Farmers and Lobster Gangs: The (un)Tragedy of the Commons

The question of just how much official regulation is needed to keep a system running smoothly is obviously a critical one for those interested in the study of cooperation. Particularly pressing for politicians, lawmakers, and social scientists is the question of how much intervention is needed to successfully regulate common property or resources—shared public spaces, municipal stores of water, and so on. The question at the heart of this debate is one that we've seen in some form or another many times before: Can humans be trusted, in the absence of strict rules and limits, to share a common resource in a way that is fair to everyone? In 1968 biologist turned ecologist Garrett Hardin published his famous parable, "The Tragedy of the Commons," the story of a village that had in its center a piece of land that was shared by all the village farmers. The problem was, there was a finite amount of grass on that commons and no laws or limits on how many cattle each farmer could allow on the land to graze. So farmers, each trying to maximize their own gain, kept letting more and more of their own cattle graze on the land, until the grass was depleted, leaving nothing for anyone. No farmer acting alone could resist this logic since the others would simply go on tragically overexploiting the commons, while the responsible farmer would simply lose out in the interim. This parable has often been cited as an argument for property rights; it seems to say (though this wasn't quite what Hardin had intended) that people are inherently too selfish to be trusted to fairly share any common resource without regulation or oversight. As it turns out, Hardin wasn't giving people enough credit. When we look around us, we see many examples showing that people are actually quite good at setting up, enforcing, and self-regulating fair arrangements—all thanks to norms.

The single most influential book to examine this problem was Nobel laureate Elinor Ostrom's *Governing the Commons* (1990), which describes more than a dozen systems of self-regulated commons throughout history that did *not* devolve into tragedy. Some of the most interesting examples are of a series of irrigation districts in Spain that had been successfully managing thousands of farmers' access to water for five centuries. With slight variations, this is how they worked. Farmers would elect officers to an informal, public tribunal that decided, based on current levels of rivers and canals, how to allocate water to each farm. In one district, in Valencia, farmers were to take turns collecting water, and when an individual's turn came, he could take as much as he needed for the duration of his turn, as long as he wasted nothing. Because farmers couldn't flood their fields, taking turns meant no one depleted the system as long as the farmers didn't sneak an extra turn (different, more stringent rules applied during droughts). How did the farmers monitor one another? (After all, a selfish farmer would be tempted to cheat the system if there were no way of the other farmers knowing he had done it.) First, they employed "ditch riders" to watch the canal and make sure no one was abusing the system. More important, there was always someone by the canals, waiting for his turn to arrive, to monitor the others. If a farmer *was* caught acting selfishly, instead of being punished by some central authority, he would simply be brought in front of a customary tribunal of other farmers.

The system worked extremely well. Even though penalties for cheating were mild, infractions were extremely rare (less than 1 percent). Violence—always a risk when such a precious resource is at stake—was practically nonexistent. It wasn't that the system was completely unregulated (very few stable, complex systems can run with no rules at all without running amok). But instead of being regulated by market-based incentives or the formal power of government, it was regulated through a set of mutually agreed upon and enforced community norms.

At this point you might be thinking, this system was set up centuries ago. Life has gotten so much more complicated since then. True. But amazingly, these Spanish systems continue to work in modern times, as do similar systems in other situations. Another of these modern examples of a flourishing, functioning commons can be found in James Acheson's study of the lobster gangs of Maine. For this collegial group of lobster fishermen, lobsters were "in the commons"; within a certain fishing area, anyone was allowed to catch as many as he wanted. But no one fisherman owned, or had any official property claim to, any of the lobster beds. Instead harbors were informally divided into sections, each shared by a group of fishermen, or a "lobster gang." The fishermen decided which gangs could fish in which part of the shoreline. In most gangs the choice spots were taken on a first-come, first-served basis, with the exception being that older, more experienced fishermen got first dibs.

In general, patrolling the boundary between groups and harbors was hardly a peaceful activity. It was enforced by a constant threat of vigilantism—a "trespasser" who oversteps the boundaries could expect to have his boats attacked or traps ruined by a member of a different harbor. Again, as I have emphasized throughout, *cooperative* does not always mean "nice." And in fact, the willingness of any given individual to risk retaliation in protecting the group's boundary (by punishing the trespasser) is itself a high form of cooperation within that person's group—the one lobster fisherman takes the risk, while everyone in the group enjoys the benefits of having their fishing rights preserved.

The day-to-day practices of both the lobster gangs and the Spanish farmers were not, in the strictest sense, "cooperative," since each individual fisherman was working to catch as many lobsters as possible for himself, and each farmer was attempting to grow the largest crop possible on his own field. But at the same time, they contributed to a common effort—such as constructing, fixing, and policing an irrigation dam, or defending a lobster fishing group's boundaries against encroachment by fishermen outside their harbor gang. And even when they were acting for themselves, they were *governing themselves* cooperatively, without the force of law hanging over them. Both the lobstermen and the farmers were able to sustain stable systems for generations without the need for explicit formal property rights and without a state exercising authority over them to prevent abuses.

Now, I'm not arguing that property rights are necessarily a bad thing. Nor am I suggesting that we just throw everything into the commons and expect everyone to take no more than their fair share. In all these studies the communities adopted some techniques to make sure that participants would curb their selfish impulses and behave well. But the point is that these rules don't have to be enforced by markets and pricing schemes, nor do they have to be imposed from above. What made the early work on commons, like Ostrom's Nobel-winning work or Acheson's on lobster gangs, so influential is that they began the long process of empirical debunking of the then-prevailing view that commons in which rules are enforced by norms are necessarily doomed to fail.

Intellectually, these studies set the background for the explosive growth in commons-based practices in the twenty-first century. For the commons has finally come into its own. Because in today's knowledge economy, the most valuable resources—information and knowledge—are themselves a public good, and the best way to develop and maximize this good is through millions of networked people pooling that knowledge and working together to create new products, ideas, and solutions. And no example better shows how successful norms can organize this very kind of cooperation than Wikipedia.

Wikipedia's Neutral Point of View

I first began studying Wikipedia in the summer of 2001, about four months after it launched. What struck me as unique about it was that, unlike all the other collaborative enterprises that were online at the time, the designers of Wikipedia didn't set any official rules or regulations; they relied entirely on a (then still simple) set of norms, and on the community, to keep the project in line. In the years since, Wikipedia has grown in influence and scope, and has occasionally been infiltrated by people who try to write or manipulate articles to further some personal, political, or commercial agenda. Yet the number of contributors and the complexity of their social system increased, and with these changes so did the complexity of the system of norms. Despite the very occasional instance of abuse (which the community is generally quick to detect and strike down), Wikipedia still offers an extraordinary example of a system that governs itself primarily through conversation and self-regulating norms rather than by market mechanisms or a formal and authoritarian management structure.

The one norm that the Wikipedia community holds closest to its heart is the neutral point of view, or NPOV. The current version of the article on NPOV in Wikipedia reads:

> **Neutral point of view** is a fundamental Wikimedia principle and a cornerstone of Wikipedia. All Wikipedia articles and other encyclopedic content must be written from a neutral point of view, representing fairly, proportionately, and as far as possible without bias, all *significant* views that have been published by reliable sources. This is non-negotiable and expected of all articles and all editors. . . .
>
> The neutral point of view is a means of dealing with conflicting perspectives on a particular topic. It requires that all majority views and significant minority views

> published by reliable sources be presented fairly.

"Reliable source" is a subjective term, especially when it comes to a topic about which people have conflicting, deeply held beliefs. Yet the NPOV norm turns out to be surprisingly effective in mediating even the most controversial and heated of debates.

Let's look at how a recent debate over an article about creationism versus evolution was peacefully resolved. In April 2008, one user, Rlsheehan (who describes himself as an engineer from Minnesota, an environmentalist, and a Bible reader), made (small, local) waves by claiming that the text of the existing article on creationism was flawed because it did not adequately acknowledge the fact that the term included a centrist view that also allowed for scientific evidence. What followed was a roughly 2,500-word exchange largely between him and another user, Hrafn (with an occasional intervention from others), over whether the opening paragraph of the article should be changed to include this third view.

My goal here, of course, is not to weigh in on who was right, but to demonstrate how norms mediated a potentially explosive argument and turned it into a civilized, productive debate that ultimately produced a cooperative solution that both sides could agree on. In a nutshell, this is how it happened. For several exchanges, the conversation centered squarely around questions of NPOV—what pieces of evidence existed, what sources counted as reliable, and what a fair presentation of conflicting views should look like. Soon, however, the tone escalated, as the debate veered into the finer points of creation, religion, and science. This is when, in response to a claim that Rlsheehan backed up by cited certain polling data, Hrafn invokes another Wikipedia norm, WP:Synth, which essentially permits authors to summarize existing materials, but only in ways that preserve their original meanings or conclusions (a subset of the norm against original research, one of the core norms of Wikipedia). Rlsheehan in turn responds by citing the NPOV norm; he claims his article is not reinterpreting information but merely representing a neutral point of view. What follows is an intense argument. Like dueling wizards, Rlsheehan and Hrafn hurl one norm after another back and forth. Eventually another user, Professor Marginalia, stepped in, cools the tone, asked for more clarity on the subject and purpose of the article, and a series of more civil exchanges ensues. By the end a solution is reached; Hrafn does a substantial edit, acknowledging that many religions accept evolution, though some are more aligned with creationists, the other with scientists. Rlsheehan deems this acceptable, and the case, for now, is closed.

Now, if what I've described—an open and spirited debate between two equals—seems an obvious way to solve a conflict, let me remind you that many of our existing systems are designed in such a way that conflicts are actually taken up a hierarchy and not resolved by following a shared set of norms. Turf wars between coworkers are generally taken straight to the boss. Settlement or custody battles between divorced couples are brought to the courts. True, mediation is a growing practice, but it is of relatively recent vintage. Even in our system of government, majority rules; the authority ends up being whoever has the most votes, and cooperation is honored only within that majority, if at all. With all these examples it is easy to forget that the basic efforts we make to teach kids to find ways to "share nicely" and accommodate one another's needs based on some shared set of norms of cooperation shouldn't end with our children. Wikipedia, for all its warts, is an amazing example of how people can work together, at large distances, not just to resolve conflicts but to solve discrete problems and produce actual outcomes—in this case a stable, reliable text. And Wikipedians do so willingly and collaboratively, using a shared set of norms, without intervention from some higher authority. Even though Rlsheehan and Hrafn may not have yielded to the other's argument, they were able to reach a solution because they continuously brought each other back to the shared set of beliefs about *how* the argument should be conducted and resolved. I don't want to overstate the utopianism of Wikipedia. It is a rapidly changing, global phenomenon that is learning and adapting before our very eyes. A new generation of scholars is exploring the emergence of various new structures of authority within Wikipedia and the complex relationship between the formal

Wikipedia foundation and the organic community itself. We will continue to learn from Wikipedia, but we cannot deny that it grew from nothing into a major global collaboration among thousands of contributors and is a system that is fundamentally collaborative and built on discussion and mutually shared norms.

So for those of us interested in designing systems that work as well as Wikipedia, the question becomes, how do we set norms that people will agree on, and follow, without regulation from above? One way would obviously be to give participants in a community free rein to set their own norms, which is partly what Wikipedia's architects did after setting the initial ground rules, including, most important, NPOV. Studies on the adoption of norms have shown that people are far more willing to follow rules and norms when they have input into their construction. In one powerful study (aptly subtitled "Self-Governance Is Possible"), Elinor Ostrom, James Walker, and Roy Gardner compared the outcomes of public goods games run under three different setups: one that let the players decide when and how to punish defectors, one that let them punish one another according to rules dictated by the experimenter, and another that let subjects communicate with one another but not punish noncooperators at all. The results were unequivocal—when participants were given the opportunity to set their own rules and choose the punishments, their levels of cooperation were highest; people cooperated as much as 90 percent of the time. By comparison, when rules and punishments were imposed by the experimenters, levels of cooperation dropped substantially, to 67 percent (although that level was still significantly higher than what anonymous public goods games normally devolve to without any kind of rules or possibilities of reward or punishment). The point is that people more readily followed game norms when they saw these norms as self-imposed, or freely chosen. As we will see when we encounter the phenomenon of crowding out in the next chapter, preserving people's sense of autonomy is crucial in harnessing cooperation, and research in self-governance suggests that this carries through to autonomy in setting the rules under which one may find oneself punished.

What this means in terms of designing systems is that even if some rules or norms must be introduced or set from above, one should still try to build in as many mechanisms for self-governance, and offer as many opportunities for people to participate in reviewing and revising these rules, as possible. Circling back to Wikipedia, we see that this is exactly what Jimmy Wales and Larry Sanger (Wales's then employee and early cofounder) did; they set an initial policy but allowed people to discuss, debate, reinterpret, and enforce it on their own.

In the 1990s, there was a surge of interest among legal scholars such as Larry Lessig and Dan Kahan in the question of just how norms can become so ingrained in a culture that people will voluntarily follow them without official regulation or enforcement. If norms are so powerful in regulating behavior, these scholars wondered, can we use law as a tool to help mold those social norms into the shape we want? Evidence suggests that this is possible, that certain laws can help reverse even some of the most long-standing norms. For example, cities that have banned smoking in public places have seen smoking rates drop dramatically. This is not just because the ban made smoking less convenient; the ban likely played an important role in making it less socially acceptable to smoke in public. As a chain-smoking tourist in New York in the late 1980s, I remember how patronizing and moralistic the antismoking ads, not to mention the nasty looks and comments from nonsmokers, seemed to me. Even after I stopped smoking, the moralizing seemed unwarranted and overbearing. But there we have it: It worked. By emphasizing the secondhand smoke effects (which countered accusations that the antismoking campaigns were paternalistic, and made them a matter of public health), the ban fostered a culture in which smoking was suddenly seen as a failure of will and self-control, rather than as a choice. In short, the law actually changed people's daily habits and behaviors to the point that businesses didn't even have to enforce the smoking ban, because people enforced it on their own. In the beginning they were following the law, but soon they were simply following the norm.

How do norms, then, become internalized in this way? There is a long line of work in psychology from Freud, through Leon Festinger, to contemporary work by John

Jost, Mahzarin Banaji, and Aaron Kay, among others, that shows how readily we'll accept and embrace given social practices and norms simply because they are already established. In one set of experiments, when students were given different predictions about the likelihood of large tuition increases, the ones who were told that the probability of an increase was high simply adapted their sense of how harmful the increase would be: They said that it would not affect them badly. On the other hand, similar students who were told that the probability was low reported that a tuition increase would be very bad for them indeed. In other words, we not only accept our reality (or what we perceive as our reality), but we also seem to trick ourselves into thinking that whatever the reality is is what we ourselves might have chosen, or is what the right state of affairs should be.

This tendency tells us a lot about why legislating a certain behavior can lead to the adoption of new norms and standards. When we construct explanations to justify why we do what we do (or in the case of smoking, not do), we eventually start to believe ourselves, and soon enough we've internalized it, consciously or not. This has significant implications when it comes to designing systems; it suggests that when we institute prosocial or cooperative norms, they become self-reinforcing over time. The systems become dynamic: The more we practice cooperation, the more we believe in the virtue of being cooperative. These changes, of course, make the problem of sustaining cooperation easier, so we can solve even more of our problems cooperatively, and so on in a virtuous cycle.

This claim is still very much a hypothesis, but it's one that is on the cutting edge of a range of disciplines. There has been some work in virtue ethics that considers (hypothetically) the effects of practice on our habits. There has also been a lot of work in social psychology about how we justify our practices, as we saw earlier. But the basic question, rooted in Aristotle's theory of how to instill virtues through practice, had not been a serious subject of study for decades. Today it is becoming a new area of research at the intersection of experimental economics, organizational sociology, political science, and evolutionary biology. Already there is enough evidence to suspect that, when it comes to cooperation, practice makes perfect—that by building and engaging cooperative systems, we increase the baseline level of cooperation throughout society.

Moral Commitments and Principled Action

Imagine you are sitting on a park bench in a city you have never been to, and to which you have no plans to return. A man walks by, hurried and harried, talking on his cellphone. He reaches into his pocket, pulls out a handkerchief, mops his forehead, and puts it back in his pocket. But in the process, he drops two fifty-dollar bills on the ground, right at your feet. He keeps walking. By now he is twenty feet away, moving fast. The park is relatively empty. There is not a soul to be seen within fifty yards of where you are sitting. What do you do? For a purely selfish person, the answer is simple. You simply lean forward and pick up and pocket the money. There are no possible repercussions. Cash is basically untraceable. There are no observers. The man is a stranger. But for a person with morals, there *are* repercussions—feelings of guilt for not having returned the money. Some of us will feel this guilt more acutely than others. Someone who has dropped money on the street themselves at some time in their life might find it more reprehensible to keep the money than someone who hasn't. Or someone who is struggling financially, while the man who lost the money appears well-off, might view it as a justified and much-needed windfall. But whatever we end up doing, if we have a conversation with ourselves, if we hesitate, let alone if we call after the man and return the money, we are acting not on fear of punishment but on an ingrained and internal moral compass.

Nothing captures our understanding of moral commitment better than the way Marx astutely put it: "These are my principles; if you don't like them, I've got others." (That's Groucho Marx, in case you didn't know.) Unlike social norms, moral codes are not so readily agreed upon, not so easily imposed or revised. This makes them powerful, where they are helpful; but they are a less malleable component of cooperative systems design.

Complicating matters further is the fact that just as

with acts driven by solidarity, not all acts driven by morals are what we would generally call cooperative (remember, a strong sense of social identity underlies many of the most horrendous acts of intergroup conflict, prejudice, and hate). Let's say you are a member of a group that has been embezzling money from the state's pension funds, but your morals get the best of you and so you turn the others in. Clearly this undermines cooperation in the group, yet most people would agree that it is the laudable and right thing to do. On the other hand, strong moral convictions can lead to quite antisocial or harmful acts, seen from a broader social perspective. An animal rights activist might vandalize a warehouse full of furs; an anti-abortion activist might shoot a physician or burn a clinic; a person committed to loyalty of friendship might withhold information about a friend who he believes is about to commit a terrible crime. The point is not that there is a single morality that drives us all. The point is that the strong impulse to act on moral principle is a general one, and that it can override quite substantial material and social interests that conflict with it.

Having a society populated by moral actors is very useful. In his paper "Rational Fools," which analyzes the importance of commitment to norms as a distinct source of action in economics, Amartya Sen demonstrated this in a delicious anecdote: " 'Where is the railway station,' he asks me. 'There,' I say, pointing at the post office, 'and would you please post this letter for me on the way?' 'Yes,' he says, determined to open the envelope and check whether it contains something valuable." The point is that if everyone constantly acted in pursuit of their own self-interest and no one actually internalized basic moral norms of honesty, fidelity, a capacity to undertake and perform commitments, and so forth, we would indeed live lives that were nasty, brutish, and paranoid. Or we would be forced to submit to a draconian system of surveillance and punishment. Of course, reality is clearly somewhere between the two extremes. Not everyone is moral; even among people of good faith who have the capacity to behave morally, not all are equally willing or able to actually do what they understand, in the abstract, is the right thing to do all of the time. But remember, what we are trying to do here is not imagine a world that is entirely different and benevolent. What we are trying to do is get a more nuanced and realistic handle on how people actually are, and to build systems that respond to this reality by allowing those people who are not driven by narrow self-interest to more effectively and efficiently cooperate with others like them without being hemmed in by the constraints of systems built on the assumption that everyone is selfish.

From a practical perspective, we can think of moral drives as particularly powerful norms. At the extreme are moral taboos, which for the most part our bodies will enforce for us. The thought of eating taboo foods (as through cannibalism) or engaging in taboo sex (like incest) literally makes us sick with disgust (though what counts as taboo in a given society can of course vary). Needless to say, not all aspects of our moral code elicit such a strong response, but it's clear that internalized moral codes do impose substantial effects on our behavior through what we might think of as our conscience.

A hot topic in psychology today is the quest to figure out precisely where and how this moral capacity works in the brain. In one study, Joshua Greene, one of the leading psychologists working on the neurobiology of morality, scanned subjects' brains while they pondered some basic moral problems (for example, whether you would flip a switch to move a trolley from a track on which five people were trapped to a track on which one person was trapped, or throw a heavy man in front of the trolley to stop it from hitting the five men, etc.). He found that, when confronted with such moral dilemmas, the parts of the brain involved not only in reasoned decision, but also in emotions, social relations, conflict, and memory—areas that wouldn't typically be involved in an objective problem or decision—lit up. Though neuroscience hasn't been able to isolate a single part of the brain involved in morality, and probably never will (the human mind is far too complex), these studies demonstrate that we do process moral decisions in physically distinct ways.

How can these physical attributes translate into the bewildering array of commitments we hold to be moral, or the wide range of taboos we practice as human beings? One possible direction was raised in chapter 2, as I discussed the work of political scientists who used twin study to argue in support of a significant genetic component to voting practices. One way for this to happen is that while we are universally born with the capacity to act morally, we must fill that capacity with actual content, as children, largely by observing others. As with the possibility that conscientiousness generally takes the form of voting specifically, so too we can speculate more generally: The specific shape that our general capacity to form and adhere to moral commitments takes depends largely on our environment and our culture. From the perspective of anyone who wants to design a cooperative system, the crucial fact is that we are capable of acting and conforming our behavior, to some extent, to the moral codes of the society we inhabit. So when we question whether someone will keep the hundred dollars dropped at his feet, or when we worry about whether someone will do a good job when they aren't being observed, or when we think about whether we need to monitor every action of a babysitter, we need to understand and remind ourselves that while we aren't all angels, most of us are wired to be moral beings.

• • •

We've seen that we are the kind of creature that cares a great deal about doing what we perceive to be right, fair, normal, and appropriate—usually even more than we care about furthering our self-interest or maximizing our rewards. This means that when we build systems, be they a municipal government program, an association's bylaws, or a new online community whose functioning depends on people following a set of norms, our approach cannot be purely about "aligning the incentives." It has to also be about helping people define what is right, what is fair, and what is appropriate in a way that is consistent with the system's goals. But this begs the question, do incentives work at all? In the next chapter, we'll look at the role of incentives in cooperation and explore the surprising dynamics of reward and punishment.

• CHAPTER 8 •

For Love or Money: Rewards, Punishments, and Motivation

It was the middle of July, in 2006. I was, as usual, on Fire Island in New York, spending the mornings writing and the afternoons building sand castles and splashing around with my kids. One morning, I was poking around on the Internet, doing research on what eventually would become this book, when I got an email from a friend, telling me to look at Nick Carr's blog. Carr is a long-standing skeptic about the social transformation wrought by the Internet, and in this particular post he was responding to a challenge by the very young and aggressive then general manager of Netscape, Jason Calacanis. Calacanis had urged the top contributors to Digg, the most popular collaborative new aggregation site at the time, to "stop being suckers" by contributing for free to Digg, and instead move to his network, Weblogs, Inc., where they would be paid for their work. Carr claimed that Calacanis's business model—to pay for content—was the future of the Internet and that while relying on volunteer, user-generated contributions, as Digg did, might be a sustainable model for now, soon, as pricing and selection mechanisms improved, the best contributions would eventually migrate to sites that paid for content.

I disagreed, so I posted a response on Carr's blog, explaining that sites that paid contributors would not be able to outcompete sites that built attractive collaboration platforms, because the motivations that drive the two—money for the former, intrinsic for the latter—are different and often in opposition. I opened my response with a flippant "I'm happy to accept this wager . . . ," and within days "the Carr-Benkler wager" had been reported in the *Guardian* and had become a Wikipedia entry. Nick and I had agreed by email that one of us would probably have to buy the other dinner sometime in 2011.

• • •

The question of how intrinsic motivations intersect with self-interest is a critical and complicated one. As I've been careful to point out throughout this book, it is one thing to challenge the prevailing view that people act only in pursuit of self-interest; it is quite another to imagine that all our actions are completely selfless. We have already seen the evidence that there are many people, besides ourselves, that we care about to varying degrees: our relatives, people in our communities, even strangers with whom we perceive some common bond or shared sense of identity. We have also seen that we care very much about doing what we believe to be right, fair, and normal, or socially appropriate, for the setting—even when those things turn out to be costly to us. But while each of these social motivations certainly factors in to our decisions about how to act, they don't tell the whole story. We are not naïve enough to assume that self-interest is not part of the story. It is only that it motivates us to a far lesser degree than we previously thought.

Perhaps the best way to see that self-interest plays a role in decision making is to go back to the lab. Economists have long operated under the assumption that the higher the stakes, the more the pursuit of self-interest guides our behavior. This turns out to be wrong. Across countless experiments of the type I've been describing, simply raising the stakes did not in fact significantly change the levels of cooperation—even when experimenters went to poorer countries, where they could make the stakes as high as three months' wages. What *did* make a difference in people's behavior was the relative cost of cooperating. When the relative gain from acting selfishly, or defecting, compared to the potential payoff for cooperating, was made smaller, more people cooperated. In other words, in a game of Prisoner's Dilemma, if the payoff when both players cooperated was say, $4, and the payoff to the defector when one defected and one player cooperated was knocked down to say, $6 instead of $9, fewer people defected. Logically, that change should not have affected the behavior if one were driven purely by self-interest. You are still better off defecting; $6 is still more than $4. But that isn't how the subjects in the experiment saw it. At least some people were willing to behave cooperatively at a personal cost to themselves, as long as the cost was not too high (in this case the cost now being $2 instead of $5). In other words, our willingness to cooperate is sensitive to the price we have to pay for the privilege of cooperating.

In the real world, one of the day-to-day practices that best exhibit this phenomenon is recycling. In cities with recycling programs, compliance is much higher when the municipalities do curbside pickup rather than requiring residents to take recyclables to some central location. This intuitively makes sense. Of course we're more likely to comply with a desired behavior when it's made more convenient. But what's more interesting is the fact that convenience turns out to be a far more important factor in determining compliance than are material incentives, in other words whether the municipalities impose fines (which seems to result in more illegal dumping rather than compliance). Other factors, such as perceived social norms (what people think their neighbors and relatives think about recycling), matter as well, but how well it works is influenced in turn by how convenient or "inexpensive" it is to recycle.

What this tells us is that if we want to encourage

cooperation, we need to keep an eye on how burdensome or less so we make it for individuals to act on their better instincts (such as when some municipalities switch to curbside recycling pickup). Almost a decade ago, when I first started studying cooperation online, I called this effect the "modularity" of contribution. In looking at online collaborative platforms, I'd discovered that the most successful ones had found ways to break the work they needed to get done into small independent modules so that volunteers could contribute in small increments in a way that was not too burdensome. Multiply those five minutes by the millions and millions of people willing to make that tiny sacrifice and there are few limits to what can be accomplished. To put it simply, it is much easier to motivate a million people to do something that will take them five minutes to complete than it is to motivate just a few people to do something that might take them months or years to accomplish.

Skeptics immediately claim, "So it is about money and incentives and self-interest after all!" If costs clearly matter, why even bother trying to harness intrinsic motivations, such as empathy, community, and a desire to do what is fair or normal? Shouldn't we just focus on rewarding desired behavior and punishing undesirable behavior—and letting the social and emotional motivations sort themselves out? The answer to this claim is simple, but surprising: We shouldn't do so, because the material and social motivations don't always work well together; in fact, rewards and punishments often crowd out, or cancel out, our intrinsic motivations.

Putting a Price on Blood

Over the past thirty years, hundreds of experiments and observational studies, primarily in psychology and economics, have demonstrated this crowding-out phenomenon. One of the first, and perhaps most groundbreaking examples, is a bloody one.

Until the early 1970s, the majority of blood donors in the United States were compensated largely in cash, which was paid by a combination of non- and for-profit entities. In Britain, on the other hand, blood donations were entirely voluntary and organized by the National Health Service. In comparing the two systems, the sociologist Richard Titmuss found that the British system had higher-quality blood (as measured by the likelihood of recipients contracting hepatitis from transfusions); less blood waste; and fewer blood shortages at hospitals (Titmuss also argued that the U.S. system was less equitable, because the rich exploited the poor and those who are desperate by buying their blood). Ethics aside, he concluded that a voluntary system was safer and more efficient than a market-based one.

Predictably, Titmuss's argument came under immediate attack from economists. Most famously, Nobel laureate Kenneth Arrow agreed that the U.S. blood system was flawed but refused to concede that it was because payments reduced the voluntary donations. Arrow admitted that some donors might be responding to moral or intrinsic incentives (giving blood because it's the right thing to do), but a completely different group of people was responding to prices and market incentives (giving blood to make money). Because these groups were separate and independent, he said, the incentives for one group had no impact on the incentives for the other. Payment should not deter the first group of givers, while removing it could only put off the second group, reducing the total number of donors. Arrow argued that the higher rate of infection in the United States could be due to the fact that donors for money were less scrupulous about donating tainted blood, while altruistic donors would not donate if they knew there was something wrong with their blood.

Despite these objections, the evidence prevailed, and the United States did in fact transition to a volunteer system in the 1970s. And when it did, altruistic donors more than made up for the slack in lost paid "donations," just as Titmuss had predicted. Once the material incentives were out of the picture, donations increased in quality and quantity and the system as a whole performed in a more stable and efficient manner than it had before the 1970s. The debate over whether paying for giving blood reduced voluntary giving remained heated in the decades that followed, but experimental results have been remarkably consistent. A very powerful recent study in Sweden, which has a background voluntary donation system, showed that women's contributions decreased significantly when they were offered payment to donate blood, although the contribution for men did not. Women's contributions returned to the original level when the women were given the opportunity to donate the money they received to a foundation that works on children's health issues.

Another study, conducted in Switzerland, offered strong evidence that introducing material rewards can reduce cooperation in other, more public settings. Experimenters asked residents of a Swiss town whether they would be willing to allow a nuclear waste dump to be built in their neighborhood. When they appealed to the residents' sense of civic duty, telling them that it was an important national goal to locate this waste safely, slightly more than half of the residents said yes. Yet when the residents were told that the parliament had voted to award monetary compensation to the residents of the town where the site would be located, only a quarter of the residents agreed—half as many as were willing to accept it when no compensation was offered! Economists such as Bruno Frey and Samuel Bowles and psychologists such as Edward Deci and Richard Ryan have analyzed similar experiments to test this phenomenon in diverse settings, across many different domains. The results at this point are fairly clear and persuasive.

But how do we explain them? Why would offering material rewards for a cooperative activity, such as donating blood, actually undermine the degree to which people participate in that activity, rather than increase it? To answer that puzzling question, imagine that our desires and behavior are like a carriage harnessed to four horses, named Material Interests, Emotional Needs (or Affective Responses), Social Motivations or Connections, and Moral Commitments. If all four horses are pulling in the same direction, we will travel in that direction—act on that preference, principle, or policy—much more readily than when the horses are pulling us in different directions. The basic idea is that there are various ways in which introducing material rewards misaligns the horses, gets them going in different directions, and therefore steers us away from wherever it is that the person offering the reward wants us to go.

The first force that misaligns these horses is what we call normative framing—or our expectations for what the norm should be in that particular situation. Think back to the Wall Street/Community Game experiments (and the examples of services like CouchSurfing.com and Zipcar) that I described in chapter 5. You'll recall that people's expectations about the type of interaction in which they are participating can have a huge impact on levels of cooperation; when they believe an interaction is social ("community"), they tend to cooperate more than when they believe the interaction is business ("Wall Street"). So when we are offered rewards for things we consider to be prosocial, or benevolent (for example, donating blood or doing a service for our country), it frames the interaction as business, signaling to us that it's therefore okay to be selfish (and vice versa).

The second misalignment mechanism is social signaling. Our behavior sends a certain kind of signal to others about the kind of person we are, and the kind of social interaction we think we are in. And we feel this sharply and intuitively in all kinds of everyday situations. Imagine, for example, that you are hosting a dinner party. One guest brings a nice bottle of wine and a box of chocolates. Another, at the end of the meal, slaps down a check for fifty dollars. I'm guessing you'd find that first friend's gesture to be thoughtful, and that second friend's behavior incredibly rude, if not downright offensive. But why? After all, they are both offering you a "payment" of sorts for the meal, and the dollar value of that payment is probably roughly the same amount. But the wine and chocolate is acceptable because it is seen as a gift or a thank-you, which is social, while the fifty dollars is seen as a business transaction, which is decidedly not social. So when we are offered payment for a benevolent activity, it signals to others that we are not as selfless as we wish to be perceived—which in many cases is at least part of our motivation to act selflessly in the first place. This is why charities, nonprofits, and arts associations often reward contributions with "gifts" like magazine subscriptions and tote bags, to frame the interaction as being more social, rather than one in which only money changes hands.

Another comprehensive and coherent psychological theory to explain the crowding-out phenomenon is one developed by psychologists Deci and Ryan. According to Deci and Ryan, people have an innate need for autonomy; we need to feel that we are in control of our own preferences, principles, and actions. So when we think we are somehow being manipulated or controlled by rewards and punishments, our sense of autonomy is threatened, and then we rebel (albeit subconsciously) by refusing to do, or by doing the opposite of, what is desired. Think about two different ways a parent might motivate a child to do well in school. The child whose parent tells him, "You did so well when you decided to work hard to prepare for this exam," will actually be more self-motivated and perform better in the long run than the child whose parents say, "I will pay you five dollars to get an A on this exam." This is not unlike what was happening with the blood donors—it was easier for people to feel good about themselves for giving because they *chose* to give than it was when they knew they were giving because someone else was offering them a payoff. If we want to avoid this crowding-out effect, then, we need to frame rewards and punishments in a way that preserves people's sense of autonomy as much as possible: Yes, they are receiving a reward, but it's a reward for something they would have chosen to do on their own.

Think back to the Swedish blood experiment. Remember that, for the women at least, when the experimenters offered money, the women reduced their donation; but when they were able to give the money to a charity, they bumped their blood donation levels back to their prior level. Here the act of publicly being able to donate the money reinforces the signal these women were sending to themselves and others that they are generous and caring. It also reframed the act as one of pure giving, instead of one that was less clearly altruistic, and allowed them to feel more in control of the interaction.

The critical point in designing systems for cooperation is that we cannot assume that offering material rewards will increase a desired behavior. Sure, it'll help spur the Material Interests horse in the direction we want it to go, but it might send the emotional, social, and moral horses tugging in the opposite direction and bring the carriage to a screeching halt.

Software Developers, University Professors, and Overpaid Executives

We live in a world where it's impossible not to care about money. We need it to feed ourselves and our children. We go to work every day to earn it. We fight wars over it. I'm not arguing that money is not important. But clearly money is more important for some people than others—that is why some of us become stockbrokers and others become schoolteachers. We know from hundreds of experimental studies that in any given situation, we can expect over half the population to behave cooperatively and generously, and about one-third of the population to behave selfishly. So what we really need are systems that harness *both* social and selfish motivations while avoiding the latter crowding the former out. In other words, we need systems that offer material rewards to those who tend to be motivated by self-interest, without putting off those who are intrinsically inclined to cooperate. We see plenty of evidence around us that this is possible, from innovative workplaces like Google, which offer a greater sense of individual achievement, community, and intellectual fulfillment but at the same time pay their employees quite well, to the numerous open-source software development projects in which many of the participants are paid to work on the project but many others are not and yet contribute for free. Let's look at some examples more closely.

"Free" as in Speech, Not as in Beer

The success of free and open-source software baffles economists no end. How can software produced largely by volunteers who do not own or reap any profits for the code they write be better than software written by paid developers? Like a bee, with its fat body and slight wings, this thing has no business flying. Yet it does. Because it is so implausible, the question of why anyone would donate their time and energy to collaborative online projects has been the subject of much research, initially in economics, and later in other disciplines ranging from computer science and process engineering to anthropology. And the answers that show up repeatedly in surveys and other studies align perfectly with all the different intrinsic motivations we have discussed so far: community, fairness, reciprocity, adherence to norms, and so on. In one classic study, Karim Lakhani and Robert Wolf found that the most common reason (cited by 44 percent of responders) to participate in open-source development was simply the enjoyment, or the pleasure of the intellectual stimulation it provides. The second most important was building one's skills. Another widely reported motivation was more normative—one-third of developers said that they thought "source code should be open." In other words, they were driven to participate because they thought it was the right thing to do. Another 28 percent cited a sense of obligation to give back to the community that had given them such great tools. Solidarity and group identity, too, played a large role; an astronomical 80 percent described the free software hacker community as their primary source of identity, and 20 percent said that they were motivated by the teamwork aspect.

So people contribute their time and effort, for free, because they think it's the right thing to do, because they think contributing is fair, because it enhances their sense of identity and community, and, quite simply, because it's fun. But let's go back to the Carr-Benkler wager for a minute. What happens when you add payment into the mix? Does it crowd out these social and emotional motivations, or does it lure the best talent away from their volunteer efforts to where the money is?

In Europe and America, about half the contributors to free and open-source projects report being paid, and that number is growing. A 2008 report by the Linux Foundation stated that more than 70 percent of work done on the Linux kernel, arguably one of the best open-source tools around, was now being done by paid coders. Does this mean the crowding-out theory is bogus? And what about our nearly universal desire for fairness? Why do those *not* being paid continue to work for free when they know others are being compensated? Exploring these two questions will go some ways toward helping us understand how we can integrate monetary incentives into a system without crowding the social ones out.

As we've seen, crowding-out effects occur when the offer of material rewards either clash with a person's expectation about an interaction (whether it's business or social) or interferes with a person's sense of autonomy and control. In a business setting, the former should not be a problem; people expect to be paid for their work. But what about control? If I pay you to do something, don't I run the risk that you will feel your autonomy is threatened and lose interest in doing the task? Not if I structure the tasks in a way that lets you maintain a level of control. One of the reasons open-source projects are so successful is that they give developers free rein over what projects to work on, what direction to take them in, and so on. Those who have studied the phenomenon find that even when a company hires a software engineer to work on a specific free software project, the company doesn't have much say in where the engineer will take the project, and even less in how the work will be used. Andrew Morton, one of the lead developers of the Linux kernel, explains that paid developers will often refuse certain requests from engineers if they don't feel the request is "good for the kernel." Most people familiar with the industry agree that if companies that pay employees to participate in free and open-source software projects didn't relinquish so much creative control to their employees, the employees would lose both interest and the social capital in the free software project itself that would make them productive employees. Both the psychological, intrinsic effects and the social effects require the employers to keep a hands-off approach with the paid free software developers they employ.

Another interesting practice that helps preserve the developers' sense of autonomy is the fact that the payment and the instructions come from different sources; the payment comes from the companies that purchase the software (like IBM), while the instructions and directions come (at least to the extent that the developers allow them) from the engineers. This is important because it seems to help the developers psychologically separate the payment from the process of creation, and the monetary incentives from the intrinsic ones. This is probably why efforts to introduce bounties—direct payments made in exchange for certain specified contributions—never took off on a large scale; they made the payment and the contribution a little too close for comfort. (While there are certainly online distributed labor markets where one can "rent a coder," these paid projects do not play an important role in free software development. They do, however, provide a place where people trained in a free software project could turn around and monetize their skills even if they were not employed by a company to develop the open-source project.) As many examples from the software world repeatedly show, looser, more indirect forms of payment that are not directly tied to specific actions are far better at motivating participation in collaborative projects. A direct appeal or a pay-for-performance model (what Nicholas Carr called Calacanis's Wallet) simply deters those who are seeking to pursue their own intrinsic interests. Models of payment, then, to the extent necessary and desirable at all, need to be more removed and distinct from the intrinsically motivated activity.

So crowding out does not occur in the open-source software world, because the industry developed practices to keep an unusual degree of autonomy in the hands of the software developers and to make the creation process a rewarding one. This all makes a good deal of sense. But that doesn't address the issue of fairness, or explain why the engineers who don't get paid don't feel slighted and refuse to participate. After all, why should half the developers get paid, and half not? Well, recall that as we saw in chapter 6, there is no fixed definition of fairness; it varies across people and situations and is heavily influenced by community and cultural expectations. Ideas about what constitutes fair payment in the open-source community seem to be very different from those held elsewhere. That's because free and open-source projects have many different kinds of currencies—recognition, status, etc.—in addition to material ones. What is absolutely clear, from my own observations as well as those of other researchers, is that in free and open-source software development communities status is based on the strength of one's abilities, on the quality and beauty of the code one has written. These intangible rewards—the respect and admiration of one's peers, the power to influence the development of the project—are valued very highly. So while the volunteers aren't being paid in the traditional sense, in this community the social rewards they receive are sufficiently valuable as to be deemed fair compensation. This has proved enough to moderate any sense of unfairness; none of the studies conducted on free and open-source projects reports tensions between paid and unpaid engineers.

• • •

Maybe you grant me that the software industry has found a sustainable relationship with the free and open-source software development communities and has managed to successfully integrate material and social motivations without the former crowding out the latter. But what if these are anomalies? What if software developers are some weird tribe with different motivations from the rest of us? Maybe the culture of free and open-source software development attracts a disproportionate number of people who, not unlike blood donors, are already inclined to value social rewards over material ones? To test this question, let's look at one of the most controversial practices of American capitalism: executive compensation.

In the 1970s, American CEOs made about 25 times more than the average worker in their firm. By the 1980s, this multiple grew to 40. But then executive pay really took off. By 1996, the ratio was 1:210, and by 2000, American CEOs made, on average, over 500 times more than average workers. The dramatic increase in executive pay was largely the result of a single idea. And the name of the idea is agency theory. Agency theory essentially says that each employee of a company is an agent working for the benefit of some other person or entity, called a principal. Thus a worker is the agent of his manager, the manager in turn is the agent of the company vice president, the vice president an agent of the CEO, and the CEO the agent of the shareholders. Now, agency theory is built around the assumption that we're all only interested in maximizing our own self-interest; agents are therefore uniformly expected to put their own goals ahead of their principal's. So, the theory goes, the best way to get the most out of an employee is to align the incentives of each agent with those of his principal. For the most part, this was easy. Each principal could set the goals and objectives for those employees who reported to them, and then pay those agents based on how well those goals and objectives were achieved. But what about the chief executives, with no one above them in the corporate hierarchy? Whose goals could theirs be aligned with? The shareholders. In 1990, economists Michael Jensen and Kevin Murphy published an influential and now classic article arguing that in order to get the most out of executives, firms needed to tie their compensation to stock performance. This lent new credibility and momentum to the practice of including stock options as a core element of executive compensation, and thus to the ballooning of executive pay to unprecedented levels (funny how academic work can become so influential when its recommendations tend to make the people who have the power to decide fabulously wealthy).

The theory worked. For executives (and perhaps the people who sold them their jets and yachts). For firms and shareholders, less so. Even Murphy, writing in 1999, conceded that it is "difficult to document that the increase in stock-based incentives has led CEOs to work harder, smarter, or more in the interest of shareholders," and in 2002, Jensen wrote that this system of compensation was encouraging executives to focus on short-term returns, which boost the value of their options, at the expense of long-term results. Countless studies that have looked for a connection between executive pay and a company's performance have not found it (case in point is the fact that Toyota, where CEO salary in the 2000s was many times lower than at General Motors, overtook GM as the largest automobile manufacturer in the world during the same period). In other words, the theory failed miserably.

But how could this be? If CEOs are rewarded based

on how the company performs, surely this would motivate them to put their best efforts into the job. Three major factors explain why it doesn't. The first is by now well-known: that when CEOs' salaries are tied to stock prices in the nearer term, they maximize short-term gains and don't emphasize the longer term. In some cases, this leads to downright fraud. One study found that companies that had stock-based compensation were twice as likely to be categorized as "fraud firms" by the Securities and Exchange Commission. The second is self-selection: Firms that pay huge stock-based compensation draw the kind of executives who are driven by monetary return, and these in turn are driven less by intrinsic rewards to do anything beyond what will affect their bottom line. The third, and possibly most pernicious, is that this model of compensation signals to everyone else in the company that money is the main currency, far exceeding effort, contribution, and talent in terms of value. One can talk until blue in the face about how the company is inclusive and collaborative, but if the people leading it are seen as selfish and overcompensated, employees will become resentful and unmotivated. Just as framing in the Wall Street game shaped the participants' expectations about acceptable behavior and action, excessively high executive salaries frame the culture of an organization as one in which it's okay to be greedy, self-serving, and uncooperative.

Still, even if tying CEO compensation to performance doesn't achieve its desired results, what other choice is there? How can corporations be expected to lure the best talent without offering competitive compensation packages? The first thing to remember is that U.S. salaries are well outside the norms for the other major economies in Europe and even more so, Japan. Money has become not an actual object in its own right, but a signal of worth among the executive's own peers; a measure of his or her value. But this isn't the case elsewhere. It doesn't take a multimillion-dollar package to attract talent to run for U.S. president, for example; the power, excitement, respect (sometimes), and other intrinsic rewards of the job are more than enough to draw many impressive candidates. Similarly, $900,000 per year (compared to tens of millions of dollars in the United States) at the Japanese automobile firms was more than enough to attract leading executives to the top positions.

Another example, near and dear to my own heart, is academia. The leading American universities are widely known for attracting the top scientific minds in the world. In today's knowledge economy, ever more dependent on innovation and technological advances, the top researchers who teach at these universities would all have much higher-paying salaries if they were to work for companies, be they Merck or Microsoft. Yet thousands of the most qualified, intelligent, creative, and educated people in the world choose to earn vastly less (in some cases hundreds of thousands of dollars a year less) in academia instead of taking their skills to the private sector. Why do they do it? Mostly because in an academic environment things like creativity, curiosity, discovery, and collaborating toward a common goal are nurtured and rewarded—not necessarily financially, but socially and intrinsically. It's no wonder, then, that leading technology companies like Google are increasingly shifting to a more campuslike environment, with much more opportunity for creativity and play, much more freedom to pursue one's own projects, and much more emphasis on engagement with the community.

In academia or in top management, in Wikipedia or in free software, in government or the nonprofit sector, we see that payment is neither a necessary nor sufficient condition to draw the best talent for work that is either inherently rewarding or widely associated in the society with respect and value. And those jobs and positions that come with well-designed social structures to support high, focused effort tend to draw exactly those people who flourish under those conditions and will go the extra mile to make the enterprise they are engaged in work.

This is why I'm so confident that sometime in 2011, Nick Carr will be buying me dinner.

The Punishment Puzzle

So far we've talked a lot about rewards. But what about the flip side, punishment? We saw in chapter 8 that people care so much about fairness, they will happily punish others for behaving selfishly, even if it comes at personal expense. But can the threat of punishment actually get people to clean up their bad behavior? Does it really have any effect, besides the satisfaction we get from revenge? To some extent we will see that the answer is yes. But we've already seen that material rewards crowd out intrinsic motivations to cooperate, so the logical question is, does the threat of punishment work any better, or is the efficacy of punishment also limited by the crowding-out effect?

Let's start with the evidence in favor of punishment. The most famous studies in this vein were conducted by economist Ernst Fehr and his collaborators, Simon Gaechter, Bettina Rockenbach, and others, using public goods game experiments. You'll recall that public goods games are experiments in which the experimenter gives subjects some number of dollars, and each participant then has an opportunity to contribute any portion of his or her endowment into a common pool. Then the experimenter multiplies the common pool by some number, and everyone shares the gains equally. Now, under this kind of setup, someone acting purely in their own self-interest will contribute nothing, letting the others contribute some of their endowment, while still sharing in the upside. But as you've probably guessed, in actual experiments, participants (on average) contribute about half what they are given by the experimenter, with many people contributing the whole and some contributing none or little (consistent with the finding that some people behave cooperatively, and others selfishly). However, when participants play multiple rounds, things begin to unravel. Once the cooperators begin to see that the selfish players are taking them for suckers, they reduce their contributions, until no one gives anything. So, Fehr and his team wondered, what if subjects had the opportunity to punish those who didn't contribute? To find out, they made it possible for participants to spend a little bit of their earnings to reduce the earnings of other participants (that is, if subject A says, "I am willing to get one dollar less for my participation in the game if you punish B" by reducing B's payoff at the end of the experiment by, say, three dollars). The results of the original experiment were unambiguous: When faced with the threat of punishment, the noncooperators began to contribute a fair share, and the system was saved.

But what about crowding out? There is extensive evidence from experiments and field observations that punishment systems can backfire at least as much as, if not more than, reward systems. One particularly instructive example comes from an experiment run in an Israeli kindergarten. The teachers in this particular kindergarten were having a problem—parents kept arriving late to pick the kids up, forcing teachers to stay past the end of their own workday. So two economists suggested to the kindergarten teachers what they thought was a great solution: imposing fines on tardy parents. One would assume this would get the parents to be on time more often, now that being late came at a price. Reality, however, was otherwise. Parents began arriving even later, on average, than before. It seemed they had come to see the fine as a fair price they paid for extra babysitting, so the guilt they initially felt about inconveniencing the teachers disappeared. Just like with the blood donations, once money entered the picture, the transaction became "business" and not "social," and so the parents felt justified in being late whenever they wanted, as long as they were paying for it. Amazingly, the change was hard to reverse. Even when the kindergarten removed the fines, the tardiness remained.

Back to the experimental drawing board, then. When Fehr and his group tested the effect of punishment using a slightly different experiment, they got surprisingly different results—much more consistent with the theory of crowding out than with their original findings. This time they introduced punishment in a trust game, rather than a public goods game. A trust game works like this: The experimenter gives each player a certain endowment, say $10. One player, the investor, then has an opportunity to transfer any portion of that back to the other player, the trustee. The experimenter then multiplies whatever was transferred by some multiple, say 3. The trustee now has an opportunity to send back however much of that amount she wants to the investor. The obvious best—and fairest—result for both of them is for the investor to invest the entire $10, at which point the experimenter would add three times that much, or $30, for a total of $40. Then the trustee should transfer half that amount, $20, back to the investor, so that each shares equally in the maximum upside. But since nothing compels the trustee to return any money, someone acting out of pure self-interest won't, and will keep the $40 for himself or herself. But at the same time, the investor knows this, so unless he is a trusting individual (or, as we saw in chapter 2, has had a nasal spray of oxytocin . . .) one would expect him to transfer nothing, and again the system unravels. If punishment worked here as it had in public goods games, then giving investors the ability to punish faithless trustees should be enough to get the investors to invest more, and the trustees to transfer more back. But here Fehr and Rockenbach were in for a surprise.

What they did was let investors tell the trustees in advance how much they expected back, *and* what the punishment, or fine, for not transferring that amount would be. Obviously this should move the selfish person to pay back exactly what was asked for, up to the amount of the fine, and ignore any requests for more, since it would be cheaper in that case to just pay the fine. What they got was something very different. Trustees sent the least amount to investors who threatened to punish them and demanded a large slice of the combined pie in the experiment. They sent even less to these threatening investors than to the investors who had no power to punish them in the experiment at all. Trustees sent the most to investors who specifically said, even though I have the power to punish you, I promise in advance not to punish you, regardless of what you send back to me. It was in response to this hyperfair treatment by investors that trustees responded with the highest transfers back to investors. This result is a classic crowding-out finding. The backlash was strongest when the investor not only threatened harsh punishment but also demanded high returns; aggressive greediness triggered the most uncooperative response.

Why did the threat of punishment work in public goods games but not in trust games? The answer, I think, lies in the different social dynamic between the games. In the trust game, the relationship is one-on-one. A threat to punish is a threat to punish *me.* When I see myself as threatened, I read what my counterpart is doing as a sign of distrust in me personally, and I am offended, and respond by refusing to do as they wish. But in a public goods game, where we all know that there are some players who are selfish and unscrupulous, suddenly the punishment is not aimed at me: It is meant for the bad apples in the group. So I'm not offended, and I respond as desired. In other words, while punishment may work well as a collective standard or norm applied to and by groups, it is a problematic tool to wield in relations between individuals.

Even in public goods games, however, experiments since the original work of Fehr and his collaborators have raised more questions about other potential negative effects of punishment. Most evocative among these have been studies looking at differences across cultures and societies. One study by Benedikt Hermann, Christian Thöni, and Simon Gaechter showed that in public goods games, people in Muscat, Oman, or Athens respond very differently to punishment than people in Boston or Melbourne. In societies that had relatively low scores on broad measures of rule of law, or harbored forgiving attitudes toward antisocial behaviors like evading taxes or evading bus fares, subjects tended to retaliate against punishers. People who would defect in the basic game also ended up—irrationally—spending some of their gains to punish people who were perfectly cooperative! In societies that had higher background rule of law measures, defection was scarce, punishment was often unnecessary, and unwarranted punishment was very rare indeed.

Of course, in the real world we have many social systems that use punishment to keep us in line, most notably the law enforcement and justice systems. Yet as the policy makers in the United States should have acknowledged by now, although we have vastly higher incarceration rates than any other comparable country, we also have much higher crime rates than countries that use more moderate and less punishing forms of public sanction. Incarceration, then, seems more like a symbolic act of revenge (or, in the eyes of some, unrequited racism) than an effective technique of reducing crime.

• • •

Where does this leave us? We know that in addition to caring about empathy and solidarity, fairness and norms, we also, at least to some extent, care about material payoffs. But we also know that people are diverse in their motivations. Not all of us are equally empathetic or community oriented; not all of us care equally about fairness and justice; nor do all of us care, to the same degree, about conforming. We also know that we are not only diverse among ourselves, but also within ourselves; the same person might be exceedingly cooperative and generous in one situation, and less so in another. Often this has a good deal to do with external circumstance. In the past, the way we have managed all these potentially conflicting motivations is by segmenting our lives into separate spheres—business, home, social life—in which we have different expectations of ourselves and of others. But this is less than ideal; in today's increasingly complex world, these spheres are constantly clashing and intersecting with one another, and the kinds of motivations that drove us in the social sphere are increasingly important to allow us to innovate and grow in the rapidly changing business environment as well. What we need is to develop new systems that incorporate both material and intrinsic motivations, in all kinds of settings—professional, personal, and social.

As we put together this new image of human motivation and the systems we need to support action by others, we need to continue to keep in mind that we are, to a substantial extent, beings of appetite, with material interests pulling us alongside, and sometimes in different directions than, emotional needs, social interactions, and moral commitments. Few if any of us are Mother Teresa. If we were, her toil in Calcutta would have seemed far less impressive. Recognizing this fact about human nature places certain limitations on how much we can lean on the various motivators when we build cooperative systems. We shouldn't build systems that require us to be saints. And reducing the costs and increasing the rewards (material and otherwise) of volunteering is an important part of building a system intended to rely on volunteerism. For online collaborations, constructing the tasks modularly—in other words, making them small enough that participating doesn't exact a huge time commitment—is critical if we want peer production to increase. The fact that someone can make a useful contribution to Wikipedia, or Yelp, or any one of the thousands of online collaborations in increments as brief as five minutes or less means that hundreds of millions of people around the globe can participate in tiny increments to produce good things. The same is true, as we saw, for recycling. This is simply about reducing the cost of cooperation. Moreover, finding ways to provide a reasonable living to people engaged in an enterprise is an important way of stabilizing their cooperation. That's why we have a thriving nonprofit sector. The idea is that there are large numbers of people in the population who have basic material needs to satisfy, but for whom money and material rewards are not in and of themselves core motivators. Building structures alongside the market to allow these people to play a significant role in providing public goods—from research to political watchdog functions—by finding ways of funding their work so that they can dedicate themselves to it is crucial. But these principles are not limited to nonmarket and nonprofit sectors. Instead, we increasingly see mixed motivational structures deployed in firms as knowledge and innovation become key and organizations must continually adapt and improve to survive. As we've seen, many of today's most successful companies operating in the most competitive markets are turning away from purely reward- and monitoring-based strategies, and instead are developing new business processes that allow for greater expression of common purpose, commitment to norms, and personal autonomy. Thus they are strengthening employees' affective commitment to the organization and its purposes. The result is that our workplace is changing: unevenly, imperfectly, but nonetheless, in some of the leading organizations, substantially and positively.

Across the board we are beginning to see increased understanding that money and material rewards are not everything, and that, indeed, their relationship to motivation and effective action is much more ambiguous and complicated than two generations of economic theory has previous claimed.

Don't get me wrong. I am not advocating that we all grow our hair and move to a commune. It's not that we should never factor material rewards and payoffs into systems of cooperation. It's that we shouldn't *only* try to motivate people by offering them material payoffs; we should also focus on harnessing their social and intellectual motivations by making cooperation social, autonomous, rewarding, and even—if we can swing it—fun. But doing so is challenging. We have to understand that adding money—as reward or as punishment—can actually redirect the total pull of our motivations and steer us off the path of cooperation.

• CHAPTER 9 •

THE BUSINESS OF COOPERATION

In 1982, the General Motors plant in Fremont, California, closed its doors. It had been slowly declining over several years. Productivity was way down. Labor-management relations were tense. On any given day, one in five employees was likely to be absent. Quality levels, unsurprisingly, had fallen well below those of the rest of the company, which itself was falling behind the market leaders in Japan. In an effort to salvage the plant, GM took desperate measures; it entered a joint venture with its rival, Toyota, called New United Motor Manufacturing, Inc. (NUMMI). Under this arrangement, GM would provide the plant and would market half the cars, while Toyota would invest $100 million and, more important (as it turned out), take over full management of the plant.

In December 1984, the factory reopened under the new management. The workforce was largely unchanged: 99 percent of the assembly line workers in the new workforce, and three quarters of the skilled workers, were UAW members who had worked at the original Fremont plant. Yet within two years the new plant had far surpassed any other GM plant in productivity and had the highest quality ratings of any automobile plant in the United States. Confidential employee surveys show that job satisfaction levels rose from 60 percent when the plant reopened in 1985 to more than 90 percent by the 1990s. The effects were lasting. Despite Toyota's recent woes, for more than twenty-five years NUMMI continued to be one of the top plants in the United States in terms of productivity and product quality. As of 2010, it is slated to become the site of a new collaboration between Toyota and Tesla, to develop electric cars.

But the success of NUMMI has had much farther-reaching implications outside of Toyota. It has forced organizational sociologists and management scientists across the country to rethink the long-standing assumption that the practices of Toyota and the other major Japanese automobile manufacturers were some quirky extension of unique Japanese culture, born out of traits so uniquely Japanese they couldn't be replicated. After all, these were American autoworkers responding to changes negotiated with an American union. The only thing that wasn't American was the management. How can we account for such an incredible and rapid turnaround? The answer is basically that Toyota Production System incorporates the very elements of a successful cooperative system that we have been examining in the past five chapters. As a result, it has been able to harness precisely the kinds of intrinsic motivations and dynamics that make workers not only more innovative and more productive, but also happier with their work and workplace.

The NUMMI plant story is simply the starkest, cleanest real-world "experiment" showing that top-performing companies—or what management science calls "high-commitment, high-performance" organizations—are those that fit the model of a functioning cooperative system. The fact is, in spite of recent safety and quality control issues at Toyota, the NUMMI example (and indeed, Toyota's long-term safety record) is clear evidence that cooperative systems don't just work for things like lowering crime and maintaining peace among lobster fishermen—they also work in business settings.

Southwest Airlines is another great example of this, as management expert Michael Beer has described in extensive detail. An oft-cited business success story, Southwest somehow succeeds in turning profits year in, year out with costs that are consistently 20 percent lower than elsewhere in the industry, despite an endemically troubled industry plagued by countless consolidations and bankruptcies, rising fuel costs, and a decline in air travel. So how do they do it? It shouldn't surprise you to hear that the answer is by allowing employees to maintain high levels of autonomy and emotional engagement; by instilling a strong sense of fairness both through its employment and compensation practices; and by fostering a strong commitment to a shared set of norms—most notably to put customer service ahead of profit. For one thing, the company makes strong relational skills a priority when hiring for all positions. It continuously emphasizes teamwork; gives individual crews in various airports the autonomy to make major decisions, such as when to turn around planes; and assigns equal responsibilities to all staff, regardless of rank, so that pilots, flight attendants, and ground crews work as peers in a way that would be considered inconceivable, socially, in other companies. Moreover, this commitment to fairness extends to employee compensation; Southwest adheres to a profit-sharing model at all levels of the company. In short, as the business studies of the company repeat, some with puzzlement: The firm focuses on relationships, rather than incentives. This is reflected even in its leadership. Herb Kelleher, Southwest's fabled CEO, doesn't just give lip service to these values; interviews with employees all suggest that they really do experience his efforts to foster a sense of community (the company's celebration of weddings and birthdays, for example) as authentic and genuine.

Examples of "high-commitment, high-performance" organizations that thrive on cooperation can be found in pretty much any industry, from Hewlett-Packard, at least until the late 1990s, to Marriott hotels, to Costco. The basic story of all these organizations is the same—each one, in its own particular business sector, with its own particular challenges and complications, made it a priority to engage employees and win their commitment to the organization through a combination of the methods I have described throughout this volume. The result? They flourished, becoming among the most innovative and profitable even in the most competitive of industries.

What Toyota Got Right

Looking at the well-studied NUMMI Fremont plant gives us a bit of an insight into how such a company looks. Now, let's remember that mass production of automobiles is a highly complex, routinized process, without a lot of room for error or variation. Yet despite the seemingly small window for change, the economists and management specialists who have extensively studied the NUMMI phenomenon have identified many ways that the Toyota's management system transformed the Fremont plant completely. The first and perhaps most wide-reaching was the change in how line employees were assigned their day-to-day tasks. In GM's Fremont, eighty industrial engineers gave each individual employee strict, detailed instructions on how tasks had to be performed, right down to every movement of an arm or pressing of a button. They timed employees, and monitored and measured their output and performance—much as Frederick Taylor had recommended a hundred years ago when he developed his Leviathan-inspired *Principles of Scientific Management.* In NUMMI, on the other hand, employees were organized into collaborative teams of four to six, each led by a team leader (to cement the solidarity even further, these were union members). There were no industrial engineers looming over every employee. Their every motion wasn't prescribed. Each team had the freedom to experiment with different ways of performing their task, and to decide, together, on the best way to perform the set of tasks that needed to be completed at their station within the time frame.

When GM was in charge, each employee had a specific, static function. Under Toyota, team members rotated jobs and were encouraged to develop an overall understanding of the entire production process. And since expanding workers' range of knowledge and job flexibility required much greater investment in job training, employees received more than five times as many training hours under NUMMI (some on company time, some on their own) as they had under General Motors.

Another big change was that the new management introduced the then unheard-of Japanese practice of *kaizen:* continuous improvement. With Toyota in charge, even the very lowest-ranking assembly-line worker could raise and vet proposals for improvements, which, if approved, became standard procedure. After a few years, Toyota introduced Problem Solving Circles, volunteer brainstorming sessions conducted informally, over lunch, which was provided by the company (never underestimate the power of free food). Managers were encouraged to be equal participants, rather than try to run the whole process, and to solicit input from team members and supervisors alike. Needless to say, at the time, this kind of employee autonomy and collaboration was highly unusual in the automobile industry, but that in itself made the system more effective; by trusting its employees in a way that none of its American counterparts would dream of, Toyota built an unprecedented level of employee engagement and trust in return.

The fundamental differences in the companies' approaches to doing business reached far beyond the factory walls. At GM, not only were employees kept on a short, controlling leash, suppliers too were treated as puppets and pitted in competitive bidding wars against one another. And just like workers and suppliers, top executives were also treated as self-interested agents—who had to be tethered to shareholder objectives through massive stock compensation schemes. At each level, the assumption was that the individual—the line employee, the supplier, all the way up to the CEO—would selfishly try to get everything they could out of the relationship with the company, and that a combination of control (Leviathan) and personal material incentives (the Invisible Hand) was necessary to keep all the pieces moving in the right direction. At Toyota (and to a high degree the other Japanese automobile companies as well), on the other hand, that same teamwork approach taken at the line level extended to both supplier relations and executive pay; they partnered with suppliers on investments, worked with them on quality improvements, and selected the suppliers based on whom they had the best long-standing relationship with, rather than from whom they could extract the lowest possible price.

(When the American automobile companies tried to replicate these more trusting relations, they behaved like addicts to the old ways—they tried to start cooperative relationships, but often when the benefits began to materialize, they would turn around and force their suppliers back into a competitive bidding process to squeeze out all the benefits of innovation that the short period of cooperation had produced. When, at a major auto show, one of the leading suppliers challenged the American companies to think about why the same exact suppliers to the U.S. plants of both Detroit and its competitors gave their best efforts to the companies that treated them better, Rick Wagoner, then CEO of GM, reputedly retorted, "Stop whining!")

The assumptions about human motivation also played out in how the companies approached executive compensation (to the benefit of American executives). At GM, the same mind-set that assumed that the lazy workers would slack off if not properly monitored and that greedy suppliers would only offer their best wares for the best price under aggressive competitive bidding processes also assumed that executives would only do right by the company if their incentives were aligned with those of the shareholders. At Toyota, as well as the other Japanese manufacturers, such as Honda or Nissan, on the other hand, it was assumed that executives, like cooperative workers and suppliers, would be motivated by intrinsic rewards as well as material ones, so there was less need for, or justification of, ridiculously inflated wages (and indeed, given what we now know about the dynamics of fairness, we'd expect wage gaps to be lower in firms driven by cooperation). At GM, CEO pay on any given year could be as much as two hundred times what an assembly-line worker would have taken home; at Toyota, it was about a tenth of that. In 2006, for example, then CEO Wagoner made more than the top twenty-one executives at Honda combined, and somewhere in the order of fifteen times the salary of his equivalent at Toyota. The difference was not, shall we say, performance based.

The end result of all this is now known. GM had to be rescued by the American government while Toyota, despite its stumbles, has emerged as the largest automobile manufacturer in the world. Of course, the growth that accompanied that rise has not been without its challenges for Toyota. As anyone who has picked up a newspaper in the past year knows, Toyota went through a very rough 2010, in particular on issues of quality. But its troubles don't negate the fact that for decades, Toyota's superior system of management resulted in cars that were more reliable and innovative, workers who were happier and more productive, and suppliers (the same exact suppliers who have such an acrimonious relationship with the Big Three) who gave them their best work and highest-quality products because they knew they were being given a square deal and a fair price. There is thus good reason to believe that Toyota will succeed in identifying where it went wrong and readjust its operations to better accommodate its new, larger scales. But even if it doesn't, few companies better demonstrate the enormous power of well-structured collaborative enterprise than the Toyota of the past three decades.

Within the field of management science, there has been a long-running debate between those who recommend controlled, "incentive-compatible" business models and those who advocate what Douglas McGregor fifty years ago called "the Human Side of Enterprise." That companies like Toyota, Southwest, and others representing the latter view are so successful could not be starker proof of the fact that success in business—any business—isn't achieved through rigid corporate hierarchies or astronomical CEO pay packages, but by fostering an inclusive, social, and collaborative workplace where performance is intrinsically rewarding.

As our world continues to flatten and the boundaries of communication continue to disappear, more and more companies are adopting these collaborative strategies. In a global economy, you never know who, somewhere in the world, will come up with a new and better way to do what you are doing. As John Hagel and John Seely Brown put it, rapid innovation is *The Only Sustainable Edge.* In this race to the top, companies that fail to look outside their corporate hierarchies and harness input and contributions from all their employees will only fall behind. The most successful know that innovation happens everywhere. Not just in executive boardrooms, or R&D labs, but everywhere—from the factory floor to the sales desk to the tarmac of an airport. These companies also know that continuous learning and innovation can't happen in an organization that treats its members like mindless robots. It can only happen in an organization that welcomes and taps the diverse insights, skills, and talents of every human being it employs. Only those organizations that have figured out how to motivate employees intrinsically, and how to engage them in the enterprise as a community of shared interest and common purpose, will thrive. As we are learning every day, that requires an organization to be human, values-driven, fair, trusting, *and* trustworthy from the top down.

Why Open-Source Works

Thanks to the Internet, today's companies and nonprofit organizations can harness the collective insights, ideas, and contributions not just of the people within the organization, but also of the millions of people outside it. Take Wikipedia, which is fast becoming the largest and most ambitious compilation of information and knowledge human culture has known.

Wikipedia is ubiquitous. Authors and journalists use it as sources. Google, which ranks results by how widely linked they are, regularly ranks Wikipedia articles at the top of just about every search result. Students usually aren't allowed to quote it in their research papers, but they will often use it as an entry point in researching a new subject. So how did it become such a global phenomenon? Let's start by comparing it to its peers, *Encyclopædia Britannica* and Microsoft's *Encarta. Britannica*'s competitive edge has long been its authority. "When you walk into a home with *Encyclopædia Britannica* on the shelf," its website says, "you know you're in a place where learning is respected." It boasts an editorial board of "Nobel laureates and Pulitzer Prize winners, the leading scholars, writers, artists, public servants, and activists who are at the top of their fields." In short, it stands for elite knowledge. For years this claim allowed *Britannica* to sell multivolume, leather-bound products for thousands of dollars. A pretty good business model, you have to admit.

Then enter *Encarta,* Microsoft's effort to capture the encyclopedia market. Like *Britannica,* it too maintained a clear separation between its knowledge creators and its knowledge consumers; it also employed paid experts, though with less distinguished pedigrees than *Britannica*'s. It then bundled this software with other Microsoft products and made it visually attractive and easy to use and navigate. In effect, Microsoft developed a mass-market version of *Britannica;* Walmart dinnerware to *Britannica*'s Wedgwood. Soon, however, both companies found themselves in competition with a business model that simply did not exist a decade ago; a model so implausible that it theoretically *could* not exist, or so we thought until just a few years ago. *Britannica* continues to soldier on, largely because its authority cannot be matched by Wikipedia. But it was forced to slash its prices. The leather-bound set now goes for $149.99, and an annual subscription to its online version costs $69.99. *Encarta* folded in 2009. The force that drove it out of the market was, of course, Wikipedia.

Wikipedia is free. That's not unusual—there are many sources of information out there that are advertiser supported and are free to consumers. Radio and television were always so in the United States, and these days, most information on the web is free as well. Much more radical than the fact that it is free to consumers is the fact that Wikipedia, unlike television and radio, doesn't pay a penny for its content; its content is produced by volunteers who write and edit it without wanting or seeking compensation, simply for the pleasure of writing, for the camaraderie of the community of Wikipedians. In short, for all the reasons we have been exploring in this book. The fruits of their collective authorship efforts is a process, not a product. A collaboration that incrementally and imperfectly improves itself over time.

In February 2001, when Jimmy Wales first came up with this crazy idea for a web platform that relied entirely on volunteer contributions, anyone who predicted that the result would one day equal or surpass the hallowed *Britannica* would have been laughed out of the room. Critics claim that Wikipedia is less accurate and authoritative than *Britannica* and other published encyclopedias. The irony is that kids (mine included) are told at school that they should not use Wikipedia for their research, while academics (on occasion, my colleagues included) will often say things like "When I want to give my students [college or graduate] a handy reference for some basic concept, I send them to Wikipedia; it's fantastic." There are good arguments on both sides of the debate, but it's hard to tell whether this mistrust of Wikipedia is warranted or merely a product of people's anxiety about this new fountainhead of knowledge. The best evidence we have toward resolving this debate is an experiment recounted in a 2005 article in *Nature* magazine, whose staff sent articles from both *Britannica* and Wikipedia to a group of leading scientists (blinded, so the scholars didn't know which ones were which) and asked the scientists to evaluate their content. As it turns out, the scientists thought that both had errors—but roughly in the same proportion. For obvious reasons, the *Britannica* staff tried to impeach the study and its methods, but those efforts largely fell flat. And for our purposes, really, the question of whether Wikipedia is just as good, slightly worse, or slightly better than *Britannica* is irrelevant; we're more interested in whether a product and platform built entirely on volunteer contribution and collaboration can succeed at all. And clearly it can, to a greater extent than anyone would have thought possible as recently as ten years ago.

Wikipedia is the most obvious example of the amazing things we can achieve when we use the Internet to harness people's collective knowledge. But it is only one of thousands. Free and open-source software like Wikipedia is an example of how a culture of open collaboration can produce an enormous amount of information. It might sound to you like the domain of geeks and hackers, but in reality if you ever visit Google, Amazon, Facebook, or the *Wall Street Journal* online, you are using free or open-source software (these sites run on GNU/Linux operating system, the Apache web server software, or both). As the founder of free software, Richard Stallman, put it, free software is not "free" as in "free beer," but as in "free speech," meaning that it is available to anyone to write and rewrite, as well as use. In the 1980s, when Stallman introduced the concept, it looked and sounded like a hippie holdover. Software should be a communal resource, open to all, was the idea, and to make it so, people could develop software and then license it to everyone under a license that allowed anyone who wanted to do so to make copies, distribute them, even sell them, all without any obligation to the original author. Licensees were even allowed to improve the software and distribute the improved software, as long as they licensed their improvements on those same open terms. It's a system that requires reciprocity and encourages continuous improvement: I give you my contributions freely; you in turn must share yours, not only with me, but with everyone else in the world who might want to use our combined creation. Stallman had nothing against people selling free software, as long as the cycle of reciprocity wasn't interrupted.

Over the next twenty years, hackers (which among software developers means cool dudes who are good at writing software, not to be confused with "crackers," who are the bad guys who most everyone else thinks of as "hackers") churned out thousands and thousands of programs, all by working in this model. In 1991, Linus Torvalds, a Finnish student, began a project to harness this model to build the kernel of an operating system. He called it Linux. In 1995, Brian Behlendorf started collecting contributions (called patches) to the university-created web-server software eventually called Apache ("a patchy") server. It became the basic server software for the majority of web servers, including the most demanding e-commerce sites in the world, for the past fifteen years. Clearly, they were on to something. By 1998 there were many software developers who were interested in "normalizing" free software and making it more mainstream. They invented a new name, "open-source software." By the end of 1998, the Linux kernel was so popular it began to look like the only serious challenge to Microsoft's monopoly on the operating system. And indeed, an internal Microsoft memo that was leaked on Halloween of 1998 confirmed that even Microsoft saw Linux as a real challenge. Whether that was what people inside Microsoft really thought, or whether it was simply a PR ruse to persuade the judges sitting in the antitrust case that the Redmond Giant really wasn't a monopoly, the fact was that there was now a serious movement afoot to persuade businesses to choose Linux over Microsoft. Suddenly, open-source software was not seen as some crazy fad, but an effective and reliable way to develop software.

By 1999 the hottest initial public offering, or IPO, in an already overheated market was that of Red Hat, whose business model at the time was selling well-packaged and maintained copies of GNU/Linux. By 2000 IBM had declared that it was investing a billion dollars in free software, and by 2003 it was making more money from selling services built around Linux than from all its patent royalties put together—even though IBM is the largest patent holder in the United States. How? The basic idea is simple. IBM is approached by a company that needs to improve its internal efficiency. To do so, it will need to combine a variety of systems to monitor the firm's work, collect and process its data, manage its communications, and so on. This requires software, but firms are complex and different, and one-size-fits-all products (like Microsoft's) often aren't good enough. But wait. What if there were a way to mix and match a bunch of off-the-shelf pieces of software and create a customized package? For this to happen, hardware would have to be sold, customized pieces of software would have to be written, and particularly well-tailored combinations of software would have to be put together—all services IBM was more than happy to offer. And better yet, what if the engineers doing the tailoring were intimately familiar with the software, and if they could freely tweak it as they wished to fit the client's needs?

This is where free software like the GNU/Linux operating system comes in. Because IBM engineers were able to collaborate with thousands of open-source developers who knew the software inside out, they were able to put together better packages, and in less time. And because the licensing terms of free software were open, those engineers had permission to tweak and tailor the software as necessary to a client's needs. What's more, because the client wasn't buying a single product like a copy of Microsoft Office, but rather a service, IBM was able to set its price according to the value of the service, not just the price of a piece of software. This is not an unusual practice for IBM. Software services actually account for more than two-thirds of the annual revenues in the software industry; less than a third is from selling the kind of software that you buy off the shelf, or by downloading. The point is that software companies are able to build a robust and profitable business model around harnessing the contributions of an open community, about half of whose members don't make any money from their contributions.

Of course, doing so requires enormous sensitivity. If a company exploits its community by failing to contribute its fair share or to respect the community dynamics, it will ultimately alienate the community and the system will fall apart. Maintaining a well-functioning, mutually supportive relationship with the community of developers outside the firm is absolutely critical to the survival and thriving of a firm that relies on inputs from a community of volunteers. One small start-up creating the first open-source video editing and hosting software, Kaltura (which later became the video content management platform for several universities, as well as for Universal and Sony Music), went so far as to hire a "C level" executive—ranking up there with the CFO (chief financial officer) or COO (chief operating officer)—to oversee community development. Collaborating transparently and in good faith, without being exploitative, is not trivial. But it is necessary.

The success of IBM and Red Hat persuaded many other firms to use open-source software, and over the course of the 2000s the number of free software developers paid for their work, directly or indirectly, rose about 50 percent. The other 50 percent remained volunteer. Yet as we saw in chapter 8, the system still worked; those who were not being paid harbored no resentment toward, and worked no less hard than, those who were. And the practice of combining paid and unpaid contributions, as well as the open, reciprocal licensing approach underlying it, has become the model for other movements, most notably the Creative Commons license that now covers millions of online works that people contribute for one another's pleasure and benefit.

Understanding how to value what people create and share online has been one of the biggest challenges for businesses in the last decade. Until very recently, "the audience" for information, delivered through traditional media like television and newspapers, was passive, and value was determined by how many "eyeballs" that information could deliver to advertisers. Even in the early days of the Internet, much content was valued according to this model (and some still is, even though it's not really accurate). But what happens when the audience is no longer passive, and the people who produce the content *are* the eyeballs? When "the people formerly known as the audience," as Jay Rosen put it, are in fact creative and intrinsically motivated to create and share their own work, knowledge, insights, and so on with one another—and then are given the platform to do it? For the "elite" creators—professional writers, journalists, photographers, etc.—this pill has been hard to swallow. But it can't be denied that the content created by unpaid "amateurs" has value. YouTube's enormous success is perhaps the most obvious example. The millions of people who post videos to YouTube every day are paid nothing for their contributions—everything from the silliest dorm-room shenanigans, to dancing pandas, to homespun music videos, to the gripping images seeping out of the Iranian reform protest movement of 2009. They draw us in. But people don't make these videos to get paid; they make them to capture their lives, to share something of the human experience—or sometimes simply to get a few laughs. And in the middle is Google, a company that has built its business selling advertisers attention and valued this breathtaking range of human video expression at $1.65 billion. It's becoming increasingly clear that the value lies not in the work of professionals, but in the attention that "the people formerly known as the audience" pay to one another's work.

Once you open up the possibility that people are not only using the web as a platform to produce their own individual content, but also to pool their efforts, knowledge, and resources without expecting any sort of payment or compensation, the possibilities for what they can create are astounding. For example, there are tens of thousands of individuals who, just for fun, spend hundreds or even thousands of hours flying virtual airplanes using flight simulators and coordinating their flights online so that they simulate real airlines, with real routes and real schedules (just run a search for "virtual airlines" and you'll see what I'm talking about). A much more meaningful example (at least in terms of how it affects people's lives) of the open-source or peer production model in action can be found in sites like PatientsLikeMe that bring people suffering from various diseases together with others battling that same disease. Some of what happens on these sites is emotional support. Much of it, however, is the pooling of observations and information about symptoms, side effects, new treatments, and breakthroughs in research, resulting in a mountain of data that can help patients, and sometimes their doctors, better treat and manage their condition. One social networking platform, Lybba, now being developed by the filmmaker Jesse Dylan (Bob Dylan's son), will even have sufficient privacy and data protections that people will be able to actually share their medical data so that it can be collected and used in actual scientific studies on their condition.

Even the U.S. military has begun to jump on the open-source bandwagon, with platforms called CompanyCommand and PlatoonLeader, which allow officers to break from the traditional silos of hierarchy to share experiences, methods, and tactics. The U.S. intelligence community has also launched Intellipedia, an internal wiki for some thirty thousand people working in the intelligence agencies. And the military in general is increasingly experimenting with what is called network-centric warfare—an approach to warfare that grants autonomy for individual officers and encourages collaboration among units—a far cry from the image of obedient soldiers marching to the orders of a distant general.

The Sounds of Music

In the past fifteen years, few industries have so completely demonstrated the triumph of the Penguin over the Leviathan than the music recording industry. When digitized music became a reality, first in the form of CDs and later on the Internet, the industry saw both the promise and the threat. The promise was what was once called by legal scholar Paul Goldstein "the Celestial Jukebox," a world where all music would be available everywhere, to everyone, for a price; a world where anytime a person listened to a song, they could be charged for it. This was a vision that presaged great profits for the music industry.

The threat, on the other hand, was that this new world would become one giant pirating machine, where music could be perfectly copied and distributed, and no one would ever pay for their music. Naturally, the Recording

Industry of America, the lobbying organization of the music industry, knew exactly in which of these two worlds it wanted to be. So, beginning in the early 1990s, it crafted a classic Leviathan strategy that used both law and technology to control the distribution and use of music and prevent illegal copying.

First, it successfully lobbied for harsher copyright infringement laws so that simple file sharing would become punishable in much the same way that large-scale commercial piracy had been before. This included a new law called the Digital Millennium Copyright Act, which made it a crime not only to copy encrypted, copyrighted materials, but also to create or distribute any technologies to help consumers to get around that encryption. Now, armed with new technological barriers as well as laws making it a crime for people to get around them—since users who tried to make copies, even legal copies, would be in violation of the new law—the industry was in a position to maintain total control over music, or so it thought.

Yet the strategy backfired culturally. While the industry thought it found a way to control music, what it had actually found a way to do was extract all the value from music. It turned out that treating music fans as freeloaders and thieves didn't usher in an era of the Celestial Jukebox at all; in fact, it only made people *less* willing than ever to pay for music.

Enter Napster, which in 1999 through 2000 broke out as the first major peer-to-peer, file-sharing network.

Kids everywhere were uploading and sharing music, for free. Piracy became cool, and hacking the encryption software even cooler. The hackers came to be seen as cultural heroes—noble warriors fighting against an overbearing industry that exploited musicians and fans alike. So what did the industry do to combat the surging piracy? It ramped up its enforcement efforts, hurling lawsuits at the creators of music-sharing sites, software developers, and even individual file-sharing fans (in one famous case even turning its legal guns on a twelve-year-old girl).

For the next ten years, the fans and the industry were locked in something of a stalemate. For each music-sharing site that the industry succeeded in shutting down, two sprung up in its place, and fans continued to share music online without paying for it (surveys suggest that the level of file-sharing activity remained more or less stable at about 25 to 30 percent of Internet users for most of the decade). Of course once iTunes and other online music stores came on the scene, the number of people paying for online music also grew, but evidence suggests this supplemented rather than curbed file sharing; as recently as 2009, industry analysts claimed that about 95 percent of online music downloads were still illegal. In short, iTunes or no iTunes, industry revenues shrunk by about 30 percent over the decade.

Clearly, the Leviathan-like strategy wasn't working. So in the late 2000s, several artists began to experiment with completely revolutionary strategies (to the "suits" in the industry, at least) that hinged not on erecting technological barriers, nor suing the pants off sixth graders, but rather on harnessing the cooperative impulses of music fans. Instead of assuming that the only way to get people to pay for music was through regulation and enforcement, these artists assumed that true fans would be intrinsically motivated to support the artists whose music they enjoyed. In perhaps the best-known example, in 2007 British rock group Radiohead released its album *In Rainbows* exclusively online and let people pick the price they wanted to pay for the download. The files were not encrypted, so there was nothing to stop fans from downloading the album for just a few cents and then making and distributing ten thousand perfect copies.

Yet, they didn't. The band didn't release figures of how well its experiment worked, but market research firms that tried to analyze the available data have estimated that two-thirds of all downloaders paid between $5 and $15 dollars. Given that across the industry 95 percent of files are downloaded illegally, and even when an album is purchased legally, under the traditional business model artists usually receive between an optimistic $1.70 per album and a more realistic 66 cents per album, the artists certainly seem to be better off with the cooperative, pay-as-you-wish model. And when Trent Reznor of Nine Inch Nails gave fans the option of downloading the regular version of his newly released *Ghosts I–IV* for free, or pay a fee to download the high-quality version, he reported that he was paid over 1.6 million dollars in the first week after its release.

When you start to think about how much this model

has in common with other cooperative systems we've been talking about throughout this book, these fans' behavior starts to make more sense. First there's the element of trust—eliminating the encryption factor and putting the option of payment entirely on the honor system. There's the element of communication, in that fans download the music directly from the artists' websites rather than from a third-party site. Then there's the strong sense of community and reciprocity; the artists encourage the fans to engage and interact with one another, and with the music (Reznor, for example, even permits fans to remix his music and make videos). And the transactions are framed as cooperative rather than market-based; the sites make clear that some payment is expected, but without policing or finger-wagging.

Still, you might be thinking, Radiohead and Nine Inch Nails are megastars—of course fans are willing to voluntarily pay for *their* music. But as it turns out, people act just as fairly and cooperatively when it comes to lesser known acts. In one study about these pay-as-you-wish models, my colleagues Leah Belsky, Byron Kahr, Max Berkelhammer and I purposely looked not at the big stars, but at what some artists are beginning to call the "new middle class" of artists—artists who are not out to make millions, but are trying to build fan bases stable enough to support them in being professional, full-time musicians.

We looked at two individual artists, Jane Siberry and Jonathan Coulton, and one online label, Magnatune, that have adopted this direct, pay-as-you-wish model and analyzed between three and five years of data on all the transactions. In each case, fans can download music that is not only unencrypted and can be copied in various high-quality formats, it's also governed under a Creative Commons license, which means it can be *legally* copied and distributed—a model that couldn't be further from the Leviathan strategies that the major labels have pursued.

But the transaction isn't the only thing that is framed as cooperative on these sites. Coulton, for example, engages his fans by making his site a place where they can not only download his work, but share and enjoy one another's as well. He not only allows but initiates direct conversations with his fans, asking them to get together to tell him where they will be so he can schedule a show there. In short, he creates a culture of empathy, reciprocity, mutual respect, and trust.

And it works. As I mentioned briefly in chapter 7, our study found that 48 percent of fans over a period of five years paid $8 per album, even though $5 was the minimum payment accepted at the site, and free copying was both technically feasible *and* legal. In fact, only 16 percent of users actually chose the $5 minimum payment, while 17 percent actually picked either $10 or $12! Coulton's revenues, for one, are such that he would have to sell hundreds of thousands of iTunes tracks every year under the traditional model to match.

So even in the music business, a troubled industry with a long history of favoring the Leviathan, cooperation can triumph. And it's all thanks to the stable, long-term relationship created between the artists who make the music—and trust their fans to pay them for it—and the loyal fans who reward them for it.

Changing the Face of Politics

Stepping outside the business arena for a moment, we see another dramatic example of how large-scale collaboration can produce amazing results in the already legendary campaign that led to the election of President Barack Obama. At the outset of the campaign, Obama was vastly outgunned by his competitor for the Democratic nomination, Hillary Clinton. She had the party establishment. She had the big donors. She had the name recognition. True, Obama had a special charisma and unique political gifts, but at the time no one dreamed that would be enough to win him the nomination, let alone the race. Many books and articles have probed, debated, and pontificated on the secrets behind his success, and though theories differ, most agree on the following: Things would have turned out differently if not for the ingenious ways in which his campaign mobilized mass collaborative action (not to mention funds) both at the local level and on the web.

This is how the grassroots campaign structure worked. A relatively small number of paid field organizers recruited local volunteers, and let each one choose a type of task, such as pulling together names of potential local supporters, or organizing community events. Those who demonstrated enthusiasm and the desire for more responsibility were promptly invited to become "neighborhood team leaders" and were put in charge of organizing local campaign activity. They were given no instructions or guidelines other than to think creatively about how to best energize the people in that unique community (after all, things that excite voters in Little Rock, Arkansas, might not be so effective in New York City's East Village, and vice versa). Neighborhood leaders, in turn, recruited other volunteers for more discrete, specific tasks—such as phone banking or canvassing door-to-door. Each volunteer recruited more volunteers. At the end of the day, more than three million people had volunteered their time, whether by canvassing streets for donations, organizing local events, calling undecideds, or driving voters to the polls. Trite as it may sound, the organizers' internal motto, "Respect. Empower. Include," wasn't just an empty slogan; each volunteer's contribution, however small, mattered.

This sense of enthusiasm, commitment, and engagement that so energized the volunteers seemed to appeal to campaign donors as well. As we saw in chapter 2, the act of voting is a puzzle to the selfishness theory, since, in a country of millions, the impact of a single vote is so negligible that even the minimal cost of voting outweighs the benefits. The same could be said of giving small campaign donations (very large donations, which can buy influence, are theoretically and practically a different matter); in a national election, the cost of making even the smallest donation would seem greater than any possible difference that that donation could make. Yet half the donations Obama received were for two hundred dollars or less—not even a drop in the bucket for what ended up being a half-billion-dollar campaign. These small donations played a particularly large role in the early stages of the primaries, when more of the traditional, bigger Democratic donors were supporting Hillary Clinton. All told, whether by sheer force of his personality or by appealing to Americans' deep desire for change, Obama inspired three million donors to open up their wallets and give for a cause they believed in.

The Obama campaign wasn't the first political movement in history to use the power of the web to harness social and political action, though it was the most effective. Howard Dean's 2004 campaign used email and online forums to organize voters. But even before that, MoveOn.org had succeeded in building the seeds of successful political mobilization using, for all practical purposes, nothing but email. By 2006, the left wing of the political blogosphere had become mature enough to essentially play the role of party press: to energize people, to enforce a level of party discipline and keep candidates on message, to do opposition research, to coordinate activists, and so on. One can say that it was actually the 2006 congressional campaign when grassroots action on the Net began to have a real and profound effect on politics. By the time Barack Obama started campaigning in earnest one short year later (though one calendar year can seem like an era in the online world), social networking had become so ubiquitous that an online strategy, once thought to be a waste of time and money, would soon become an integral part of every serious campaign. Social networking—Facebook in particular—proved to be a crucial tool for Obama. Chris Hughes, one of the cofounders of Facebook, came on board early on to spearhead the social networking effort. His brainchild, MyBarackObama.com, was a social community, just like Facebook, that allowed Obama supporters to build profiles, post on one another's walls, write messages, and organize offline events (without direction or instruction from the campaign staff), as well as create their own mini-online campaigns—set targets, invite friends and family to donate, and show how close they were coming to their personal fund-raising goal. The site was a huge success; by the end of the campaign, it had over two million users spread throughout the fifty states, and had posted over two hundred thousand campaign events. Through creative new approaches like the personal small campaigns, or matching new donors to current donors whose profiles they could see, the site made the interactions seem more personal and helped the users feel like they were really connecting with a larger group. This was a big part of what made the site the hub of the Obama campaign's breathtakingly successful online fund-raising effort.

By now, many details of this story of how Facebook helped put Obama in office are well-known. But a lesser-known backstory exists. Around 2002, the political blogosphere began to evolve into a legitimate source of information and political commentary. Again, this is known. But what is less known is how the blogosphere was utilized differently across the political divide. In a recent study, Aaron Shaw and I looked at the top 155 political blogs of the time and found that the left- and right-wing blogs were actually quite different in terms of the amount of commentary, debate, and discussion they invited and allowed. The left-wing blogs, like Daily Kos, used technologies that allowed for multiple authors (software systems called Scoop, Drupal, SoapBlox, and ExpressionEngine) and included plug-ins (for common platforms, like WordPress or Blogger) that allowed any user to post unlimited and unrestricted comments. The right-wing blogs, like Instapundit, were more often authored and controlled by a single person, and rarely integrated contributions from anyone but a select group of main authors. As a result, the right-wing blogs became platforms for amplifying the existing discussion, keeping the role of readers relatively passive, while the left-wing blogs became platforms for actual debate, discussion, and cooperation. This wasn't a trivial distinction. It seems to have had a real impact on shaping the landscape for the 2008 election, starting with the 2006 midterm election, when the Daily Kos (founded and edited by Markos Moulitsas Zúniga), the most visible blog on the left (receiving about as many daily hits as the major media outlets), became a major force in raising awareness, funds, and votes for those Democratic candidates whom the major party institutions thought too peripheral to support (even pressuring senators and representatives from safe seats to transfer money from their war chests to marginal contenders). What this did, in effect, was prove that the playing field was open to lesser-known candidates—like Obama—who didn't have the giant coffers or influential backing of the presumed front-runners.

What the Obama campaign did brilliantly was to marry two radically separate communities—on-the-ground community organizers and left-wing bloggers—into a single network of activists, volunteers, and donors. For the ground organizers, creating that human connection and sense of community that would motivate people to contribute was relatively easy. There were plenty of opportunities for the local volunteers to meet face-to-face, and they felt a natural sense of solidarity simply by virtue of being like-minded people from the same region who were passionate about the same issues and working toward the same goal. What was more impressive was how MyBarackObama.com was able to replicate this among a vast network of strangers dispersed throughout the country, in large part by giving people the autonomy to call upon whichever volunteers they chose, organize their own events, and so forth. This approach differed markedly from the standard model political consultants had long adopted, and it was rewarded with literally millions of volunteers.

As we look at the campaign in hindsight, we can see that the Obama campaign is to political campaign strategy what Toyota was to the automobile industry, or what Wikipedia and free software are to the business of information. It is a shining example of just how effective we can be when we stop looking at human behavior as being motivated by self-interest and instead embrace a more nuanced, complex view of what drives us as human beings.

• CHAPTER 10 •

How to Raise a Penguin

It turns out not to be the case that "if you wish to build a society in which individuals cooperate generously and unselfishly towards a common good, you can expect little help from biological nature." It turns out that we are much messier and more complicated than that; nature made us that way. It turns out that evolutionary biology in the past two decades has developed ever more refined models to explain natural cooperation: from simple direct reciprocity, to indirect reciprocity, networks and affiliation, and ultimately group or multilevel selection. It turns out that our brains light up differently when we cooperate with other humans; it makes us, at least many of us, happier. It turns out that we can and do trust, and behave in a trustworthy manner. Not all of us. Not all of the time. But much more so than we have long allowed ourselves to assume when we tried to solve large, complex problems of human cooperation, whether managing a company, writing an encyclopedia, or running a political campaign. Now, at the end of the first decade of the twenty-first century, we are poised to apply these lessons to improving the systems in which we live, work, and play.

We are not angels. But we are also not the benighted, self-interested automatons that those models based purely on the Invisible Hand or Leviathan would have us believe we are. We have a variety of needs, goals, and motivations; we are all indeed concerned with material self-interest, at least some of the time, and at least to some extent. But many of us are not *so* concerned with self-interest that we permit it to overwhelm the many other things we care about. We are diverse in our motivations. Some of us care more about one set of needs or goals than others; sometimes we change what we care about over time. And all of us care differently about different things, in different situations. We care about others and our relationships with them. And we care about what we see as right, fair, and normal. These are the kinds of beings we are, naturally, socially, and culturally. That is why those of us who want to design a successful system—be it a new law, a business model, a web platform, or a local volunteer effort—need to be cognizant of the kinds of cooperative beings and society we in fact are. If we wish to build a society, an organization, or a technical system in which individuals cooperate, then we need to build these systems to account for all these motivations, and for the complex interactions among them.

Designing for Cooperation

As we try to aggregate and understand the wealth of research coming from so many disciplines about human motivation and cooperation, it helps to isolate certain "design levers" or elements of successful cooperative human systems that we can employ to motivate the people working within such a system to contribute to the collective effort rather than exclusively pursue their own interests (at the expense of those of the group). Think of these design levers as a universal translator: from the language of research and science to the language of practical application. When I list these levers here, I don't mean to imply that they are equally appropriate or even available for all activities, for all types of cooperative systems. Different activities or different populations are better served by different combinations of these levers. A campaign to get people to donate funds for disaster relief, for example, is better off appealing to a shared set of moral commitments and sheer human empathy than is a telemarketing sales department trying to get its employees to deliver better service. But even the telemarketing department that understands the importance of fair wages and employee autonomy will do better than one that uses technology to strictly monitor the work environment, or one that implements a reward system based exclusively on material incentives. So, with that caveat in mind, what follow are the levers I believe, based on the evidence I've cited throughout this book, to be ingredients of successful, practical, cooperative systems.

Communication. Nothing is more important in a cooperative system than communication among the participants. As we've seen throughout this book, when people are able to communicate, they are more empathetic, more trusting, and can reach solutions more readily than when they cannot talk to one another. Communication is key to the system's success.

Framing, fit, and authenticity. As we've seen, people react differently to situations depending on how those situations are framed (think of the Wall Street/Community Game). But people are also not stupid. If you try to frame an obviously competitive or exploitative system as a collaborative one, you might fool some of the people some of the time, but you will not be able to fool enough people all the time to make the system work. It is important that the frame in fact *fit* the reality. So while framing a practice or system as collaborative, or as a "community," may encourage cooperation for a while, if that claim isn't authentic and believable the cooperation won't last.

Looking beyond ourselves: empathy and solidarity. Again, as we've seen, face-to-face communications and letting people truly get to know one another (the more meaningful the level, the better) had actual, measurable effects on cooperation. For reasons both biological and social, the more empathy and solidarity we feel with others, the more likely we are to take their interests into account. Just as feeling empathy or solidarity with other individuals makes us more disposed toward cooperating with them, solidarity with a group makes us more likely to sacrifice our own interest to the collective whole. Of course, this is a complex and potentially dangerous road, as the difference between "in-group" solidarity and "out-group" discrimination is a slippery slope. When it comes to solidarity it's also worth noting that different people respond to different cues, and there are indications that there may even be gender differences in the degree to which solidarity matters. In short, while it's impossible to ignore or deny the role of team identity, or team spirit, in getting people to cooperate, when building it into a system we do need to be wary of its potentially dangerous effects.

Constructing moral systems: fairness, morality, and social norms. Whether you're designing a business model, a website, or a legal statute, values are not an afterthought. Fairness is not something you attend to after the practical decisions about how to improve efficiency or innovation or productivity have been made. Fairness is integral to effective human cooperation. We care about fairness, and when we believe that the systems we inhabit treat us fairly, we are willing to cooperate more effectively.

This is possibly the most critical finding for purposes of legal and organizational design. For decades, mainstream economists have persuaded legislators and judges, business leaders and management consultants that writing a good law or setting up a good organizational structure requires first and foremost "getting the incentives right." But what the research on cooperation has reveled repeatedly is that people care about the fairness of their interaction independently, and no less powerfully, than about the pure payoff from the interaction. In short, fair systems are productive systems. What's more, "getting incentives right"—that is, perfectly aligning rewards with performance—is a lot trickier than it sounds. Given that it is nearly impossible to have a system that can perfectly monitor and reward or punish every aspect of performance we care about, we need to rely on intrinsic motivation instead. And to be intrinsically motivated, people have to believe that the system they are working in is fair: that its outcomes are fair, that its processes are fair, and that others who have influence over them in that system intend to treat them fairly. Fairness is a precondition to productive collaboration.

We also care about morality; by this I don't mean we are prudes. We care about seeing ourselves as people who do the right thing, whatever our understanding of what "the right thing" to do is. Clearly defined values are also crucial to cooperation; quite simply, discussing, explaining, and reinforcing what the right or ethical thing to do in a given setting is will increase the degree to which people behave in that way. This is not the same thing as having a formal ethics code with set penalties for violations. As we saw in the example of the Israeli kindergarten parents, who were more likely to break the rules when there was a financial penalty, punishment can often have the opposite of the intended effect. In addition, morality, like fairness, is impossible to enforce perfectly. If we didn't have deeply engrained moral prohibitions against harming others, we would never be able to peacefully coexist; there simply aren't, and can't be, enough police officers or jailers to keep us in line.

But while moral commitments of this sort (against murder, for example) are relatively stable and universal, others vary across situations, contexts, time periods, and cultures. That's why it's important for a cooperative system to have a set of codes that are less predicated on "rules" than on social norms, and that are more malleable over time. Moreover, they must be transparent; letting people in a system or interaction see what others are doing reinforces the social norms and gets people to further comply with them—not because they fear embarrassment or ostracism, but because they want to do what is "normal." Remember the tax experiments I described in chapter 7, or the music downloading sites to which thousands of fans chose to contribute what they perceived of as the normal amount, without actually being forced to do so at all. Articulating social mores can go a long way into getting everyone to align their behavior *without* the need for rewards, punishment, or monitoring and control.

Reward and punishment. For all the evidence that we care about others, or about doing what's right and fair, we still care about ourselves. Most of us, perhaps to differing degrees, care about our own material payoffs at some time or another (and some of us all the time). So rewarding desired behavior and punishing undesirable behavior can, in some instances, be effective in getting people to work together and achieve a collective goal. And even though the main thrust of this book has argued that humans are motivated by far more than merely carrots and sticks, there are times when rewards and punishments, if administered properly, can work. The challenge is to find a way to motivate those who are self-interested to cooperate, without losing those who are more socially and intrinsically driven. No person designing a human system, cooperative or otherwise, can afford to completely ignore material motivations. The decision one has to make, however, is which ones to harness, and how. One thing that seems to be quite clear from the experimental literature, and the many examples of "crowding out" I've cited, is that rewards are likely to be more effective than punishment, because punishment can result in resentment or even retribution, which can undermine long-term cooperation. While our criminal system shows that punishment can sometimes work to keep people in line, it is also a very delicate tool that can misfire. Rewards, on the other hand, don't create the same alienation and resentment, but they do run the risk of drawing in people who are simply self-interested and reward-seeking, and repelling people who are interested in cooperating for more intrinsic reasons.

In the business world, companies and organizations approach the question of rewards and incentives very differently, depending on the type of organization and how sensitive the people within it are to social and moral motivations. Socially oriented organizations and nonprofits will normally offer fewer economic incentives (in other words, lower pay) but make up for it with social and moral rewards (being seen by others as a good person, and the internal satisfaction of doing good). Academia is similar in that what it lacks in economic rewards it makes up for in social and intellectual ones. Wall Street firms do the opposite, harnessing the self-interest of employees with the interest of the firm by tying very high payments to performance. In that sort of setting, such an incentive scheme can work, but only if the results are precisely and easily measurable; the gains from cooperating are low; and there is a direct, stable correlation between what each individual does and achieves and what the collective effort achieves. If, however, the activity involves greater risk taking or innovation, or requires cooperation among people whose respective contributions are hard to monitor and measure—like know-how, intuition, insight, or going the extra mile—then an incentive-based system will likely draw the wrong people and be unlikely to work. So when designing a compensation scheme, it's important to look at what rewards are most likely to be important to the types of people that the organization or venture is likely to attract.

Reputation, transparency, and reciprocity. Cooperation hinges upon long-term reciprocity, both direct and indirect.

As the Ben Franklin example in chapter 2 makes clear, systems that rely on reciprocity, particularly the indirect or "pay-it-forward" kind, are enormously valuable but also easily corrupted, or "invaded by knaves." Reputation is the most powerful tool we have against this. As online systems like eBay have shown us, even reputation systems that are essentially anonymous—such as "handles" that betray nothing of a person's actual identity—can be enough to keep people in line

Building for diversity. The fundamental claim I am making is that we have diverse motivational profiles—we differ from one another in how we respond to different motivational settings. But we are also diversely motivated individually—that is, each of us responds to a range of motivational drivers that don't sum up simply, and don't always carry the same weight in all contexts all the time. Because we differ from one another and we are diverse, systems that seek to harness us to collective efforts or encourage us to coordinate and cooperate with others have to be somewhat flexible. They also need to recognize that we are sensitive to the cost of cooperating, yet the degree of our sensitivity can fluctuate over time. A system that depends on massive self-sacrifice is perhaps not impossible but is extremely difficult to sustain in the long term. The fate of the major nationalist and communist experiments of the twentieth century in Germany, Russia, and China provide ample evidence of that fact. There's only so long we can expect people to sacrifice their own interest for the collective good.

But recognizing that we all have, in varying measures, a healthy dose of self-interest doesn't mean we are selfish automatons, either. Systems that harness our diverse motivations are not just more productive but also more consistent with the human experience than those built only for people who care about material payoffs and letting the rest sort themselves out. One of the best ways to do this is by allowing for *asymmetric contributions*—in other words, letting some people contribute a lot and others relatively little. In order for this to work, however, people within the system have to overcome the mind-set that asymmetric contributions amount to a free ride. Sure, there are some cases where some people do much of the work and others seem to share in equal benefits. But this often isn't quite as it seems. If someone makes enormous efforts on Wikipedia or on a free software development project, they may not be paid any more than anyone else (if at all) or receive any greater share of the end product, but they are rewarded in other, more social ways—such as being perceived as a leader, or expert. Even with the 80:20 rule, which says that 80 percent of the effort or output comes from 20 percent of the people, the best way to get that remaining 20 percent out of the other 80 percent of people is to allow them to contribute small amounts—whether it's of time, effort, insight, or money—while making the 20 percent high contributors feel generous, valued, and good about their contribution (and not taken advantage of by "free riders"). Encouraging those 80 percent of people to make even the tiniest of contributions—like those charity drives that set up shop on the street corners of large cities and urge passersby to give just a penny (while of course allowing those who wish to give more to do so)—are the best way to maximize contributions, and cooperation, across the board. This model of letting people contribute as much or as little as they wish has also been the hallmark of successful online collaborative platforms.

• • •

For centuries we have been trying to understand how to build human systems for the kind of people we really are. We keep oscillating between philosophical approaches that assume universal selfishness, like Thomas Hobbes's *Leviathan,* and systems that assume a basic human benevolence, anchored in Rousseau. We have even seen sophisticated observers trying to hold on to both visions of human nature and behavior, such as Enlightenment scholars David Hume and Adam Smith, who believed us to be moral creatures yet nonetheless also believed that our economies and systems of government needed to be anchored in the assumption of universal self-interest. Over the centuries, many have harnessed science to this debate; Herbert Spencer adapted Darwin to his support of laissez-faire, and Peter Kropotkin showed evolution to favor cooperation. Modern science—from biology to the study of human behavior—continues this tradition, and we are today at a moment where the study of cooperation is on the rise. Over the past twenty years we have seen an increasing body of work dedicated to bringing this science into practice. Leaders in fields ranging from sociology and psychology to management and computer sciences have been exploring how to apply this science to build systems that will best harness our diverse motivations for the common good.

We have fought over how to build human systems for centuries because our motivations are so fundamentally diverse. We know ourselves to be selfish, but we also know ourselves to be generous, fair-minded, and decent. We know that we are imperfectly both. We've seen ourselves and others behave admirably, and we have seen others (and not always quite so clearly, ourselves) act selfishly.

Against the background of this diversity, a very powerful working assumption has been to build systems aimed at the bad people, the people out for themselves, and let the others sort themselves out as they will. David Hume wrote: "In contriving a system of government . . . every man ought to be supposed to be a knave and to have no other end, in all his actions, than private interest." Or as Justice Oliver Wendell Holmes, Jr., put it 250 years later, "If you want to know the law and nothing else, you must look at it as a bad man, who cares only for the material consequences which such knowledge enables him to predict, not as a good one, who finds his reasons for conduct, whether inside the law or outside of it, in the vaguer sanctions of conscience." Building laws and governmental structures, business processes and technical systems aimed at Holmes's "bad man" has always seemed less risky. At least, we reasoned, the worst won't happen.

Always aiming to constrain the bad man might be the safer option. It might even be perfectly rational. But it also makes us always miss out on what would happen if we did trust. In life, we take chances on one another. We trust, and we behave in trustworthy ways. Not always; not with everyone. But much more often than the cynical and unflattering views of human nature and interaction would predict. And when we do, it turns out that we thrive; at the least we do better than when we do not trust anyone.

It is that larger truth that I have tried to bring to light in this book. I've hoped to show, by cutting through the breadth of scientific and observational evidence now available to us, that we aren't suckers or naïve idealists when we trust, or reciprocate trust. And along the way I've hoped to show how cooperation trumps self-interest—maybe not all the time, for everyone, but far more consistently than we've long thought.

The past fifty years have been dedicated to refining systems design based on this narrow, crabbed view of human nature. Let's dedicate the next fifty years to the vastly more complex but infinitely more rewarding task of designing the systems we inhabit for the kind of diverse, complex, but overall fair-minded, moral, sociable, and humane beings we in fact are.

ACKNOWLEDGMENTS

The origins of this book are in two conversations I had about five years ago, around the time that my previous book came out. That book had capped a decade's work on the emergence of the networked information environment. A major claim I had made in my work leading up to and in that book was that large-scale cooperation, such as free and open-source software or Wikipedia, was not a bizarre side story of the Net, but a core vector through which the transition to a networked society and economy was happening. The first conversation was with Vera Franz, a program officer at the Open Society Institute, who had asked me whether I could help the OSI think about how to set up an open-access science publication system in Central and Eastern Europe. The problem was exactly the kind of problem I was studying online—the construction of an online cooperation system. The difficulty they were facing was that, at least at that point in time, it wasn't clear how to get the scientists to contribute their work to the open-access publication system. As I talked with her about it, I realized that my work, and that of others looking at online cooperation, had very much been focused simply on overcoming disbelief. We had all functioned within an academic and policy system so completely enamored with the idea that self-interested rationality ruled the human-behavior roost, that my work was focused overwhelmingly on explaining why online cooperation was happening, why it was a stable feature of this new environment, and why it was central to the future of networked society. What I hadn't spent anywhere near as much time doing was learning how to answer Vera's question: What do I need to do to get *this particular* cooperative venture going. This was, for me, the point at which I turned from focusing exclusively on the macro-level explanations of why online cooperation was happening as a broad social phenomenon, to the question of how to design cooperative systems. The second conversation was with Silicon Valley serial entrepreneur Tara Lemmey. We were sitting in Helsinki; it was February. Tara asked me why I didn't bother trying to write for a broader audience. My book was dense; my articles, denser still. I said I liked it that way—it was my element, I knew how to do it, plus there were plenty of other people who could write books more accessible to the public, and everyone should recognize what they are good at. Tara looked at me and said: "It's just a skill, like any other; you can learn new stuff, you do it all the time. Learn this too." And then she made the strongest argument of all. I had just released *The Wealth of Networks* under a creative commons noncommercial license, because I wanted to make sure no one was excluded from getting it because they couldn't afford to buy it. "There's more than one kind of barrier to access," Tara noted; density and academic writing norms are no less of a barrier than copyright. They just exclude different people in different ways. So it was from the combination of these two conversations that this book came: an effort to explore the design of cooperative human systems, from the ground up, in a form aimed at a reader willing to work at a book, but not at cutting through forms essential to academic rigor, and that get in the way of readability. If you do in fact find that this book is more readable than my last one, then you, and I, have Talia Krohn and Roger Scholl, my editors at Random House/Crown, to thank. If not, then my own obstinacy is to blame.

I have been enormously fortunate in the students and fellows I have worked with in developing the research underlying this book. The first group of students was a research group I had set up when I was teaching at Yale Law School.

The student with whom I worked for the longest time while at Yale and who made the largest systematic contributions to my work on cooperation was David Tannenbaum, who worked on the experimental and theoretical economics pieces of the research. The group included fantastic work from Shyam Balganesh, Sarah Faulkner, Anne Huang, Josh Rolnick, and Patrick Toomey. A long-term project with Leah Belsky and Byron Kahr—both students of mine at Yale Law School—and Max Berkelhammer, an earth sciences Ph.D. at USC and, with Byron, a band member in City of Progress, was utterly fascinating and ended up in the voluntary donation in music study that plays an important role in chapter 7.

When I moved to Harvard in late 2007, I was lucky enough to have Dave Rand and Anna Dreber Almenberg walk into my office to talk about online cooperation. Dave was a Ph.D. student in Martin Nowak's Program on Evolutionary Dynamics at Harvard; Anna an economics Ph.D. at the Stockholm School of Economics visiting PED. Instead of one interesting conversation, what developed from there was a set of research collaborations and a three-year seminar on cooperation at Harvard's Berkman Center for Internet and Society, collecting a set of brilliant students and fellows, both from Harvard and from other universities in Boston. We had people from evolutionary biology, computer science, psychology, sociology, law, business, education, anthropology, all showing up and spending hours just translating to one another and trying to understand things like "What do you actually *mean* when you say 'cooperation' or 'altruism'?" It was an exhilarating experience. In addition to Dave, who was the animating spirit, and Anna, this group included many from whom I learned much; among them were Kobi Gal, Mako Hill, Andres Monroy-Hernandez, Thomas Pfeiffer, and Fiery Cushman.

• • •

A major part of my work, in particular the experimental side and the construction of online experimental platforms, over the past few years has been supported by the Kauffman Foundation's grant to the Law Lab at the Berkman Center; the Ford Foundation's support was central to my observational work on cooperation both online and in industrial settings. I am indebted to Bob Litan for the former, and to my co-investigators, John Clippinger and Oliver Goodenough, on the Kauffman grant. I am very much in the debt of Leonardo Burlamaqui of Ford for his support to the cooperation project. A central pillar of my research group on cooperation has been Aaron Shaw, a sociology Ph.D. student at Berkeley, a research fellow at the Berkman Center, and a collaborator for several years now. He managed a research group looking at forms and models of online cooperation, which had excellent participation and contributions from Yael Granot, Anna Kim, Tim Hwang, Roxanna Myhrum, Ayelet Oz, and Dharmishta Rood. On the industrial cooperation side of the work, Carolina Rossini was the powerhouse who organized a network of researchers and developed and wrote the major research reports from that work. In these, and even more so in the many ongoing research collaborations that have not made it into this book, I am indebted to Laura Miyakawa, who has been a simply unparalleled project manager and made it possible for me to be effective despite working on more projects than I could ever keep track of.

• • •

As I worked over the past few years trying to refine and synthesize the diverse materials and fields, I have benefited from being able to present various versions of the argument in many places, and to benefit from the insights of colleagues. An inspiration throughout this work has been Sam Bowles, who first invited me to a multidisciplinary workshop on the coevolution of behavior and institutions at the Santa Fe Institute in 2004, and inspired me to start looking at the range of disciplines that have ultimately come to populate this book. My first public presentation of the overall framework of cooperative human systems design was at the Hawaii International Conference on System Sciences in January 2008. The person who was responsible for my invitation there, and whose wisdom and insight have been an invaluable guide for many years, was John Seely Brown—a sage with a subversive twinkle in his eye if ever I met one. I was fortunate enough to participate in a working group on peer-production and systems design that was organized by Duncan Watts and Micheal Kearns, and that provided many insights to my work here. I got an important boost to thinking about the applications of cooperation to the legal system design side of things from presenting at the Tobin Project in 2008, and am grateful to David Moss for including me in the whiteboard project there. In 2009, I was lucky enough to work with David Parkes to put together a Radcliffe Institute exploratory seminar on cooperation and human-systems design, trying to bring together people working no computational systems and computational mechanism design, his field, and the new work on cooperation. That group, including Mahzarin Banaji, Iris Bohnet, Amy Bruckman, Yiling Chen, Joe Konstan, Pete Richerson, Charles Sabel, Luis von Ahn, and several of the students and fellows I have already mentioned produced a fascinating multidisciplinary feast. More recently a talk in Zurich gave me the opportunity to get invaluable comments from Bruno Frey and Stephan Bechtold. There was a certain delight of closure in that the last talk I gave before this text was done was again in Santa Fe, at the SFI's public lecture, where I tried to present publicly this mixture of accessibility to an engaged, thoughtful nonacademic audience, while trying to not introduce too many errors on the natural and behavioral sciences side of things. The audience stayed to the end and didn't nod off: that's one good sign; and the scientists didn't seem to find too much to complain about either, which also may be a good sign. I was particularly fortunate to spend time with David Krakauer after the talk, who helped me think through and refine some of the points on evolutionary biology, and with Murray Gell-Mann, who didn't seem to think I was bullshitting more than most. We'll see.

INDEX